中等职业学校教学用书（计算机技术专业）

计算机实用英语
（第2版）

主　编　林延葵　韦树梅　李晓霞
副主编　唐海霞　秦　妮　郑荣强　龙初英
参　编　唐立芳　雷春玲　唐　磊　高丹夏

电子工业出版社

Publishing House of Electronics Industry
北京·BEIJING

内 容 简 介

本书是中等职业学校计算机技术专业的配套教材。本书根据中等职业学校课程改革的要求，体现了文化课如何更好地为专业服务。全书主要以计算机专业知识的学习中用英文方式呈现的内容节选为教材内容，共 4 篇 12 章，即基础篇、硬件篇、软件篇及网络篇，旨在为专业教学做引领。同时，本书也是学生将来学习到相关专业内容时的一本工具用书、参考资料。本书根据"轻语法、重词句"的指导思想，以"够用"为原则，分门别类地指导学生学习计算机操作过程中接触到的英语词句，力求让学生在平时专业学习时能了解掌握常用的专业应用英语，激发学生学习计算机英语的兴趣，增强其学习信心，并使其在掌握专业知识的基础上更好地学习英语，在学习英语的同时进一步巩固专业知识。

本书内容实用性强，图文并茂，浅显易懂。考虑到书中涉及计算机专业的教学内容，为方便老师教学及未接触过计算机专业相关内容的学生更好地学习，本书还配备了相应的教学课件，详见前言。

本书适合中等职业学校计算机技术专业及其他相关专业使用，也可作为各类计算机培训的教学用书。

未经许可，不得以任何方式复制或抄袭本书之部分或全部内容。
版权所有，侵权必究。

图书在版编目（CIP）数据

计算机实用英语 / 林延葵，韦树梅，李晓霞主编. —2 版. —北京：电子工业出版社，2010.8
中等职业学校教学用书. 计算机技术专业
ISBN 978-7-121-11357-4

Ⅰ. ①计… Ⅱ. ①林… ②韦… ③李… Ⅲ. ①电子计算机—英语—专业学校—教材 Ⅳ. ①H31

中国版本图书馆 CIP 数据核字（2010）第 133900 号

策划编辑：关雅莉
责任编辑：谭丽莎
印　　刷：三河市鑫金马印装有限公司
装　　订：三河市鑫金马印装有限公司
出版发行：电子工业出版社
　　　　　北京市海淀区万寿路 173 信箱　邮编　100036
开　　本：787×1 092　1/16　印张：11.5　字数：294 千字
版　　次：2007 年 8 月第 1 版
　　　　　2010 年 8 月第 2 版
印　　次：2023 年 8 月第 17 次印刷
定　　价：28.00 元

凡所购买电子工业出版社图书有缺损问题，请向购买书店调换。若书店售缺，请与本社发行部联系，联系及邮购电话：(010) 88254888，88258888。
质量投诉请发邮件至 zlts@phei.com.cn，盗版侵权举报请发邮件至 dbqq@phei.com.cn。
本书咨询联系方式：(010) 88254617，luomn@phei.com.cn。

前言

计算机技术日新月异，新概念、新术语、新资料源源不断地从国外引入，直接采用英文术语（或缩略语）的现象越来越普遍；伴随 Internet 应用的日益普及，网上涌现了大量的英文信息；计算机操作过程中所出现的菜单、提示、帮助及错误反馈信息也常以英文界面出现，若不能迅速理解其含义，将影响我们的学习、工作和生活。

鉴于上述情况，更重要的是对中职英语教学过程中呈现的问题的反思，我们编写了本书。在组织和编写过程中，我们对比以往的教材做了较大的调整和改进，力求体现中职学生的特点和专业需求，并为教师引入新的教学思路和教学方法。

本书具有以下特点：第一，内容的组织按计算机专业的专业主干课程进行分类，便于学生在进行专业英语而非专业相关知识的学习中有一个清晰的思路；第二，全书以计算机专业课程教学中用英文方式呈现的内容节选为教材内容，目的是为专业教学做引领并成为专业学习的参考资料；第三，本书多以屏幕截图方式显示相关计算机操作过程中的菜单、提示、帮助及反馈信息的英文界面，便于学生学习和操作；第四，本书以对专业名词及词汇的掌握为侧重点，以利于学生的拓展和应用，可达到服务于专业的目的。

为了便于任课教师教学，本书提供了如下教学参考。

一、目的与任务

本课程是中等职业学校计算机技术专业及相关专业的一门专业课程，其主要任务是使学生掌握计算机专业教学的主要内容和计算机基本操作中常见的英语词句，懂得实际操作及常用软件中以英文描述的内容的含义，具备识别常见屏幕英语且能根据提示进行相关操作，能使用以英文方式呈现的常用软件的能力，为专业服务。

二、教学内容

第一部分为基础篇，共 3 章，主要介绍计算机发展过程中的一些标志性事件和人物、IT 业内的知名人物和著名的公司。

第二部分为硬件篇，共 3 章，主要介绍常见的硬件和设备名、计算机的启动、BIOS 设置、计算机系统的安装与卸载。

第三部分为软件篇，共 3 章，主要介绍常见的操作系统、常用的软件和一些工具类软件。

第四部分为网络篇，共 3 章，主要介绍计算机网络的基本术语和常用的网络硬件、Internet 的应用。

附录中主要介绍如基本商务中常用的英语词句和电子商务的专业术语、计算机常用的翻译工具和计算机词汇的特征，并收录了本书各章节出现的主要单词及计算机应用常见的词汇。

三、教学基本要求

计算机实用英语是计算机技术专业及相关专业的学生应具备的基本专业知识。通过学习，学生应能熟知并能应用书本及网络中常出现的英文名词，能识别常见的屏幕英语，会使用以英文方式呈现的软件，以达到在日常的计算机学习和应用中，不再因英文表述而受到严重影响的目的。

四、与其他课程的关系

不管是计算机的操作，还是计算机其他专业课程的学习，其内容采用英文术语（或缩略语）的现象越来越普遍，操作过程中所出现的菜单、提示、帮助及错误反馈信息也常以英文界面出现，部分常用软件没有合适的中文版或没有完全汉化，这都要求学生掌握基本的计算机英语，具备初级的专业英语能力。因此，计算机实用英语是学好其他专业课程的基础。

五、学时分配：60 学时

学时分配建议见授课学时分配表。

授课学时分配表

章　节	课　程　内　容	学　时
1	The History of Computer	2
2	Leading Figures of IT Industry	2
3	Famous Computer Companies	4
4	Computer Knowledge	2
5	BIOS Settings	8
6	Installing and Uninstalling	8
7	Operating System	2
8	Application Software	6
9	Tool Software	7
10	Computer Network	2
11	About Internet	3
12	On-line Chatting and Entertainment	4
附录 A	商务篇	4
附录 B	常用的翻译工具	2
附录 C	计算机词汇的特征	4
附录 D	单词表	
合　计		60

本书由桂林市职业教育中心学校林延葵老师、柳州市第一职业技术学校韦树梅老师和南宁市第一职业学校李晓霞老师担任主编，负责本书编写大纲的制定、组织编写和审核。

为使该书在确保技术可操作性的同时更贴近中等职业学校的教学实际，本书的编写队伍均由中等职业学校的一线教师组成。他们是唐海霞、秦妮、郑荣强、雷春玲、唐立芳老师。其中本书的第 1、11 章和附录 B 由桂林市职业教育中心学校秦妮老师编写；第 2、3 章由广西理工学校唐海霞老师编写；第 4、5、6 章由桂林市职业教育中心学校郑荣强老师编写；第 7、8、9 章由李晓霞老师编写；第 10、12 章和附录 A 由韦树梅老师编写。

根据各使用该教材的学校反馈的修改意见，以及多方征求相关课程老师的建议，并得到了来自各方的指导和支持，本书的编写组对 2007 年出版的《计算机实用英语》的原教材进行认真审视后做了改编。本书保持原来的编写初衷和思路不变，但对其内容进行了较大的变动。为了方便教师教学，本书还配有教学指南和习题答案，有需要的读者可登录华信教育资源网免费注册后下载。

感谢读者一直以来的支持和鼓励，让我们有机会做得更好。

由于作者水平有限，书中难免有疏漏与不足之处，敬请广大读者和任课老师提出更多、更好的建议和改进措施。

编　者

2010.6

CONTENTS
目 录

第一篇 基 础 篇

CHAPTER 01　The History of Computer （计算机历史） ··· 3
  Exercises ·· 5
  参考译文 ·· 5

CHAPTER 02　Leading Figures of IT Industry（IT 业内的知名人物） ····························· 7
  Part Ⅰ　About Bill Gates ·· 8
  Part Ⅱ　Do you know of them? ··· 9
  Exercises ··· 10
  参考译文 ··· 10

CHAPTER 03　Famous Computer Companies（IT 业内的知名公司） ··························· 11
  Part Ⅰ　Famous Computer Companies ··· 13
  Part Ⅱ　Hardware Brands ·· 14
  Exercises ··· 16
  参考译文 ··· 16

第二篇 硬 件 篇

CHAPTER 04　Computer Knowledge（计算机基础） ·· 19
 Lesson 1　Commonly Used Hardware （常见的硬件设备） ·· 20
  Part Ⅰ　PC ··· 20
  Part Ⅱ　Keyboard ·· 21
  Part Ⅲ　Commonly Used Hardware ·· 22
 Lesson 2　Booting Computer （计算机的启动） ··· 22
  Exercises ··· 24
  参考译文 ··· 25

CHAPTER 05　BIOS Settings（BIOS 设置） ··· 27
  Part Ⅰ　AMI BIOS Software ··· 29
  Part Ⅱ　Main ·· 29

    Part Ⅲ Advanced ·· 31
    Part Ⅳ Power ··· 32
    Part Ⅴ Boot ··· 33
    Part Ⅵ Exit ·· 34
  Exercises ··· 36
  参考译文 ··· 36

CHAPTER 06 Installing and Uninstalling（安装与卸载）··· 37

  Lesson 1 Installing Operating System（系统安装）·· 38
    Part Ⅰ Installing Windows XP ·· 39
    Part Ⅱ Ghost ·· 40
  Lesson 2 Installing Device Drivers（安装设备驱动程序）··· 46
  Lesson 3 Installing Applications（安装应用软件）··· 49
    Part Ⅰ Download Adobe Reader 9.3 ·· 49
    Part Ⅱ Install Adobe Reader 9.3 ·· 50
  Lesson 4 Uninstalling Applications（卸载应用程序）··· 53
    Part Ⅰ Using control panel ··· 54
    Part Ⅱ Using Uninstalling program ·· 55
  Exercises ··· 60
  参考译文 ··· 60

第三篇 软 件 篇

CHAPTER 07 Operating System（操作系统）·· 63

    Part Ⅰ Windows——The most popular operating system ····························· 64
    Part Ⅱ Mac OS——A operating system only good for graphics/ media work ······ 65
    Part Ⅲ Linux——A free operating system ·· 66
    Part Ⅳ UNIX——A different operating system ·· 67
  Exercises ··· 68
  参考译文 ··· 68

CHAPTER 08 Application Software（应用软件）·· 71

  Lesson 1 Microsoft Office（微软办公室配套软件）··· 72
    Part Ⅰ An Introduction to Functions ·· 72
    Part Ⅱ Application of Function ··· 73
    Part Ⅲ Commonly Used Functions in Excel ··· 75
  Lesson 2 Macromedia Dreamweaver（Macromedia 公司的 Dreamweaver）························ 76
    Part Ⅰ Commonly Used HTML Tag ·· 76
    Part Ⅱ Dreamweaver tag inspector ·· 77
    Part Ⅲ Dreamweaver HTML Code ·· 78
  Lesson 3 Macromedia Flash（Macromedia 公司的 Flash 软件）······································ 81

　　　　Part Ⅰ　Familiar Word in Flash ··· 81
　　　　Part Ⅱ　ActionScript ·· 82
　　　　Part Ⅲ　Familiar ActionScript code in Flash ································ 85
　　Exercises ··· 85
　　参考译文 ··· 86

CHAPTER 09　Tool Software（常用工具软件） ································· 87

　Lesson 1　Compression and Decompression Software　（解压缩软件） ········ 88
　　　　Part Ⅰ　Interface of WinRAR ··· 88
　　　　Part Ⅱ　How to use WinRAR ··· 88
　Lesson 2　Picture Browsing Software　（图片浏览软件） ······················ 90
　Lesson 3　AntiVirus Software　（杀毒软件） ······································ 92
　　　　Part Ⅰ　Virus ·· 92
　　　　Part Ⅱ　Scan Viruses ··· 93
　　　　Part Ⅲ　AntiVirus LiveUpdate ··· 95
　Lesson 4　Firewall——Agnitum Outpost Firewall（防火墙） ················· 97
　Lesson 5　Screen Capture Software　（屏幕捕捉软件） ························ 98
　Lesson 6　Burning Software　（刻录软件） ······································· 100
　　　　Part Ⅰ　Interface of Nero ··· 100
　　　　Part Ⅱ　How to use the Nero ··· 101
　　Exercises ··· 103
　　参考译文 ··· 103

第四篇　网　　络　　篇

CHAPTER 10　Computer Network（计算机网络） ····························· 107

　Lesson 1　Basic Terminology　（基本术语） ····································· 108
　　　　Part Ⅰ　Basic Terminology of Network ·· 109
　　　　Part Ⅱ　Common Terminology of Internet ···································· 109
　Lesson 2　Common Kinds of Network Hardware（常用网络硬件） ········ 110
　　Exercises ··· 112
　　参考译文 ··· 113

CHAPTER 11　About Internet（关于Internet） ·································· 115

　Lesson 1　Primary Internet Services　（Internet的主要服务） ················ 116
　　　　Part Ⅰ　Information Query and Publish ·· 116
　　　　Part Ⅱ　Communication for Information ······································ 118
　　　　Part Ⅲ　Resource Sharing ··· 120
　Lesson 2　Search Engine（搜索引擎） ··· 121
　　Exercises ··· 125
　　参考译文 ··· 126

CHAPTER 12　On-line Chatting and Entertainment（网上聊天与娱乐） ……………… 127
　　Lesson 1　Instant Messaging　（即时通信） ……………………………………… 128
　　　　Part Ⅰ　ICQ ……………………………………………………………………… 129
　　　　Part Ⅱ　MSN & QQ …………………………………………………………… 134
　　Lesson 2　Entertainment on the Internet（网上娱乐） …………………………… 137
　　　　Part Ⅰ　Games on the Internet ………………………………………………… 138
　　　　Part Ⅱ　Shopping On the Internet …………………………………………… 139
　　　　Part Ⅲ　Web TV ………………………………………………………………… 140
　　Exercises …………………………………………………………………………………… 140
　　参考译文 ………………………………………………………………………………… 142
APPENDIX A　商务篇 ……………………………………………………………………… 143
APPENDIX B　常用的翻译工具 …………………………………………………………… 149
APPENDIX C　计算机词汇的特征 ………………………………………………………… 159
APPENDIX D　单词表 ……………………………………………………………………… 165
参考文献 ……………………………………………………………………………………… 173

第一篇

基础篇

CHAPTER 01

The History of Computer
（计算机历史）

The first digital computer, known as ENIAC, was first formally put to use on Feb 15th, 1946 at the University of Pennsylvania in the USA. It contained 17468 vacuum tubes, consumed 174 kilowatts, covered an area of 170 square meters, weighed 30 tons, and could perform addition at the rate of 5000 times per second. At the time, it was the absolute champion of calculation speed, and its computing precision was also unprecedented.

Decades after the birth of ENIAC, many replacements have taken place inside the computer. Electronic devices have evolved from vacuum tubes, to transistors, to small-scale, medium-scale, and large-scale integrated circuits, to ultra-large scale integrated circuits. At present, computer applications have been extended to all areas of society.①

Interrelated Knowledge

公元前 3000 年，古埃及开始结绳记录，这是人类记数的开端；而六七百年前，中国发明了算盘，它被认为是最早的数字计算机。进入中世纪，欧洲人陆续发明了加法机、乘法机、差分机等机械式计算工具，为近现代电子计算机工具的进一步发展打下了坚实的基础。1904 年，英国人弗莱明发明了真空电子二极管。电子管的诞生，是人类电子文明的起点。1906 年，美国人德弗雷斯特（无线电之父）发明了电子三极管，从而促成了无线电通信技术的迅速发展。1938 年，德国科学家朱斯制造出 Z-1 计算机，这是第一台采用二进制的计算机；1943 年，英国科学家研制成功世界首台可编程电子计算机"巨人"计算机，专门用于破译德军密码。

1944～1945 年间，美籍匈牙利科学家冯·诺伊曼在第一台现代计算机 ENIAC 尚未问世时提出一个新机型 EDVAC 的设计方案，其中提到了两个设想：采用二进制和"存储程序"。这两个设想对于现代计算机至关重要，也使冯·诺伊曼成为"现代电子计算机之父"，冯·诺伊曼机体系延续至今。

Words and Expressions

digital	['didʒitl]	adj. 数字的，数码的，手指的，电子的
formally	['fɔːməli]	adv. 正式地，形式上地
vacuum tubes		真空电子管
consume	[kən'sjuːm]	v. 消耗，花费，挥霍
calculate	['kælkjuleit]	v. 计算，估计，核算，计划，认为
absolute	['æbsəluːt]	adj. 绝对的，完全的
champion	['tʃæmpjən]	n. 冠军，优胜者　vt. 保卫，拥护
precision	[pri'siʒən]	n. 精确度，准确（性）　adj. 精确的；准确的
unprecedented	[ʌn'presidəntid]	adj. 空前的，前所未有的
primarily	[prai'mərili]	adv. 首先，主要地
transistor	[træn'zistə]	n. 晶体管
small-scale		小规模
medium-scale		中规模
large-scale		大规模
ultra-large scale integrated circuits		超大规模集成电路
replacement	[ri'pleismənt]	n. 更换，接替者
abacus	['æbəkəs]	n. 算盘
super	['sjuːpə]	adj. 超级的，极好的　n. 主管人，负责人

表 1-1　计算机的变迁

算盘（abacus）	加法机（adding machine）
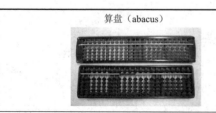	
Abacuses can do addition, subtraction, multiplication, division.②	Frenchman Pascal in the 17th century caused a sensation for creation of the world's first mechanical adding machine in France.③

续表

巨人计算机（Colossus）  During the second world war British scientists invented computer "Colossus". "Colossus" made great contribution to speed up to the end of the second world war.④	"银河"巨型计算机（"银河" super computer）  December 22, 1983, the first super computer named "银河" of China which can do more than 100 million operations per second has been successfully developed in Changsha.⑤

ENIAC——Electronic Numerical Integrator And Computer （电子数值积分计算机的缩写）	

Exercises

1. Match.
   1）computer   A. 操作
   2）operations   B. 逻辑
   3）Network   C. 电路
   4）circuit   D. 网络
   5）logician   E. 计算机

2. Thinking.
你认为在计算机发展史上最重要的事件是什么，为什么？

参 考 译 文

① 世界上第一台电子数字式计算机于 1946 年 2 月 15 日在美国宾夕法尼亚大学正式投入运行，它的名称叫 ENIAC（埃尼阿克）。它使用了 17468 个真空电子管，耗电 174kW，占地

170m^2，重达 30t，每秒可进行 5000 次加法运算——在当时它已是运算速度的绝对冠军，并且其运算的精确度和准确度也是史无前例的。

ENIAC 诞生后短短的几十年间，计算机的发展突飞猛进。主要电子器件相继使用了真空电子管、晶体管、中、小规模集成电路和大规模、超大规模集成电路，引起计算机的几次更新换代。目前，计算机的应用已扩展到社会的各个领域。

② 算盘能够进行加、减、乘、除运算。

③ 法国人帕斯卡于 17 世纪制造出世界上第一台机械式加法机，在法国引起了轰动。

④ 第二次世界大战期间，英国科学家研制出了计算机"巨人"。"巨人"为改变"二战"进程立下了汗马功劳。

⑤ 1983 年 12 月 22 日，中国第一台每秒运算一亿次以上的"银河"巨型计算机在长沙研制成功。

CHAPTER 02

Leading Figures of IT Industry
（IT 业内的知名人物）

If the IT industry is compared to one large country, the well-known people of IT would be the country's heroes. They demonstrate their skills to create various marvels, while leaving behind inspiring stories. You can surf the net (using search engines such as Baidu) to learn about the deeds of these heroes.[①]

Interrelated Knowledge

IT 业及其发展前景的几大关键词

IT 技术：IT 的英文是 Information Technology，即信息技术。它主要包括计算机硬件和计算机软件两个层次，其中计算机软件是 IT 技术的核心。

市场巨大：政府政务更加透明，城市管理更加先进、快捷，因此需要完善金融、财税、海关、经贸等信息系统。IT 技术在电子商务、电子娱乐、远程教育和学校上网等方面都有涉及。

发展迅速：十年以来，特别是国务院 2000 年 18 号文件实施以来，中国软件产业一直在以非常快的速度发展，每年大概都保持 30%左右的一种增速，而且据预计在未来的很长一段时间里，中国的软件产业仍将继续保持很快的一个速度在发展。

人才奇缺：全球软件人才存在大量缺口，欧美、日本、印度等国家和地区均面临着软件人才的短缺问题。中国每年软件人才需求缺口为 40 万～100 万人次。大量高等院校毕业生面临择业困境，而大批 IT 企业却难以招收到适用的人才。

待遇丰厚：据国家权威信息收集部门调查，全国软件工程师平均起薪为 3000 元，入职两年以上的软件工程师年薪达到了 6 万元以上。

Words and Expressions

leading	['li:diŋ]	adj. 领导的，主要的，在前的
figure	['figə]	n. 图形，数字，形状；人物，外形，体型
information	[ˌinfə'meiʃən]	n. 信息，情报，新闻，资料，询问
technology	[tek'nɔlədʒi]	n. 技术，工艺（学）
industry	['indəstri]	n. 工业，产业，勤勉
entrepreneur	['ɔntrəprə'nə:]	n. 企业家，主办者，承包商
revolution	[ˌrevə'lu:ʃən]	n. 革命，旋转，转数
corporation	[ˌkɔ:pə'reiʃən]	n. 公司，法人，集团
found	[faund]	v. 建立，创立，创办
provider	[prə'vaidə]	n. 供应者，赡养者
executive	[ig'zekjutiv]	n. 执行者，主管，行政部门
chairman	['tʃɛəmən]	n. 主席，会长 vt. 担任……的主席（议长）
title	['taitl]	n. 头衔，名称，标题，所有权，资格，冠军
Oracle	['ɔ:rəkl]	Oracle 是世界领先的信息管理软件开发商；殷墟出土的甲骨文（oracle bone inscriptions）的英文翻译的第一个单词

Part Ⅰ About Bill Gates

1. Keyword：PC Windows Microsoft William (Bill) H. Gates

William (Bill) H. Gates is one of the best-known entrepreneurs of the personal computer revolution. In 1975, he founded Microsoft Corporation with Paul Allen. Today, Microsoft has become the world's largest provider of computer software. Bill Gates stepped down as executive chairman of Microsoft on June 27, 2008, but he remains at Microsoft as non-executive chairman.[②]

2. Learn more about Bill Gates (**http://www.thegatesnotes.com**)

"盖茨笔记"引导页如图 2-1 所示。

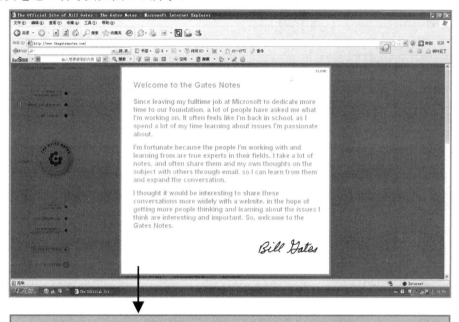

图 2-1 "盖茨笔记"引导页

Part Ⅱ Do you know of them？

IT 风云人物如表 2-1 所示。

表 2-1 IT 风云人物

Name	Title	Remark
斯蒂夫·鲍尔默 （Steve Ballmer）	CEO, Microsoft	他们是 eWEEK 2008 年年度 100 名 IT 领军人物排行榜前 20 名、美国知名 IT 网站 CRN 评出的 2009 年 IT 市场最具影响力的 25 大高管之一
拉里·埃里森 （Larry Ellison）	Co-founder & CEO, Oracle Corporation	
迈克尔·戴尔 （Michael Dell）	founder, CEO, and chairman, Dell Inc.	
斯蒂夫·乔布斯 （Steve Jobs）	Co-founder, Chairman and CEO, Apple Inc.	
王　安	文字处理机之父、王安公司创始人	全球 IT 业较具影响力的华人
施振荣	中国台湾最大 IT 公司宏碁（Acer）的创始人、主席	
杨致远	全球最大门户 Yahoo!的创始人	
王　选	当代毕升、方正董事长，中国软件技术的象征	

Exercises

1. Thinking.

（1）IT 业可分为哪几大发展领域？你喜欢的 IT 领域是什么？你将如何规划你的 IT 生涯？

（2）你最崇拜的 IT 英雄是谁？他（或他们）在哪方面作出了杰出的贡献？他（或他们）最值得你学习的地方是什么？

2. Understand the meaning of the following abbreviations and its full name in English.

IT

CEO

参 考 译 文

① 如果把 IT 业比喻成江湖，那么 IT 业的著名人物就是江湖中的各路英雄，他们各显神通，为 IT 这片天空创造出了各种奇迹，并留下了激励人心的故事。建议同学们通过网络（如通过百度）来搜索和了解这些英雄的事迹。

② 威廉·亨利·比尔·盖茨三世是个人计算机革命最知名的企业家之一。1975 年他和保罗·艾伦一起创建了微软公司。今天，微软已经成为世界上最大的计算机软件提供商。2008年 6 月 27 日，比尔·盖茨辞去了微软执行董事长一职，但他仍担任微软的非执行董事长。

③ 欢迎来到"盖茨笔记"网站

自从辞去了在微软的全职工作而投身于慈善事业，很多人会问我在做什么。很多时候我感觉自己又重回了校园，将大部分的时间用来学习我感兴趣的事。

我很幸运，因为和我亦师亦友的这些人在他们各自的领域都是真正的专家。我经常做笔录，并通过邮件分享他们和我自己的见解，这样就能从他们那里汲取知识并让交流更加深入。

为了能知道更多人在这些我感兴趣并认为重要的问题上的看法和感受，我想如果通过一个网站将这种交流扩大化，那么将有更大的意义。所以，欢迎来到盖茨笔记。

比尔·盖茨

CHAPTER 03

Famous Computer Companies
（IT 业内的知名公司）

With the development of PC, the IT enterprise gains a good opportunity to develop. In order to DIY a PC, knowledge about a variety of computer hardware manufacturers and how to choose the monitor, CPU, mainboard, etc. is needed. In order to install software for work or entertainment, it is essential to compare different brands of software from different developers. Now, let's learn about some famous IT companies and hardware brands.

Interrelated Knowledge

PC 是 Personal Computer 的缩写，是指能独立运行、完成特定功能的个人计算机。个人计算机可以在不共享其他计算机的处理器、磁盘和打印机等资源的情况下独立工作。它可分为桌面型计算机和笔记本计算机两大机型。

PC 若按系统分，可分为国际商用机器公司（IBM）集成制定的 IBM PC/AT 系统，以及苹果电脑所开发的麦金塔系统。目前通常说的 PC 是指 IBM PC/AT 的兼容机种。Apple 机与 PC 的主要区别在于它使用的是苹果机专有的操作系统。

微型计算机的普及与广泛应用，应归功于 Apple 计算机（苹果电脑公司的创始人乔布斯成立公司后的产品）的发明，以及 IBM 公司出品的 PC。

Words and Expressions

famous	['feiməs]	adj. 著名的，一流的
company	['kʌmpəni]	n. 同伴，客人，一群，连队，公司
personal	['pə:sənl]	adj. 私人的，个人的
compatible	[kəm'pætəbl]	adj. 能共处的，可并立的，适合的，兼容的
logo	['ləugəu]	n. 图形，商标，标识语
flash	[flæʃ]	n. 闪光，闪现，计算机动画技术
premiere	['premiɛə]	v. 初次公演，初演主角
illustrator	['iləstreitə]	n. 插图画家
reader	['ri:də]	n. 读者，读物，读本
design	[di'zain]	n. 设计，图样；vt. 想象，设计，计划
integrated	['intigreitid]	adj. 整合的，综合的，集成的
advanced	[əd'v:nst]	adj. 高级的，先进的
device	[di'vais]	n. 装置，设计，策略，设备
international	[ˌintə'næʃənəl]	adj. 国际的，世界性的　n. 国际比赛
business	['biznis]	n. 商业，生意，事务
Internet	['intə:net]	n. 因特网，国际互联网
electric	[i'lektrik]	adj. 电的，令人激动的，鲜亮的
printer	['printə]	n. 打印机，印刷工
mainboard	['meinbɔ:d]	n. 主机板，主板
memory	['meməri]	n. 记忆，内存，回忆，【计算机】存储
mouse	[maus]	n. 老鼠，鼠标，胆小如鼠的人
keyboard	['ki:bɔ:d]	n. 键盘；vt. 用键盘输入
monitor	['mɔnitə]	n. 班长，【计算机】显示器，监视器
		v. 监视，监听，监督

Part Ⅰ　Famous Computer Companies

世界著名的计算机公司如表 3-1 所示。

表 3-1　世界著名的计算机公司

Name	Logo	Introduction
Microsoft	Microsoft	微软，有时缩略为 MS，是全球最著名的软件商，美国软件巨头微软公司的名字。"Microsoft"一词由"microcomputer"和"software"两部分组成。其中，"microcomputer"的来源是"微型计算机"，而"soft"则是"software"软件的缩写。该名字是由 Bill Gates 命名的。顾名思义，微软（Microsoft）是专门生产软件的公司。当今 90％以上的微机都装载有 Microsoft 的操作系统，如 MS-DOS 6.22，Windows 3.2，Windows 2000，Windows XP，Windows Vista 等
Adobe	Adobe	奥多比，是全球最大、最多元化的软件公司之一。该公司的 Adobe Photoshop 是最受欢迎的强大图像处理软件之一。2005 年，Adobe 公司以 34 亿美元的价格收购了原先最大的竞争对手 Macromedia 公司，这一收购极大丰富了 Adobe 的产品线，提高了其在多媒体和网络出版业的能力。Adobe 的流行产品有 Adobe Photoshop，Adobe Dreamweaver，Adobe Flash，Adobe Premiere，Adobe Illustrator，Adobe InDesign，Adobe After Effects，Adobe Reader 等
Intel	intel	英特尔，商标由"INTegrated Electronics（集成电子）"两个单词的缩写而成，是世界上最大的 CPU（中央处理器，被人们称为计算机的心脏）及相关芯片的制造商。80％左右的计算机都使用了 Intel 公司生产的 CPU。其产品从早期的 8088 发展到曾经令人们称赞不已的 Pentium（奔腾），再发展到目前流行的 Core（酷睿）等
Apple		苹果，美国苹果电脑公司，该公司以生产高性能专业级计算机著称于世，与一般的计算机不同，苹果机的操作系统是 Mac OS X
AMD	AMD	超威，AMD 的全称为 Advanced Micro Device 。按照英文直接翻译叫"高级微型设备"公司。一般将其简称为"超威"。它是世界第二大 CPU 制造商，产品有 K5，K6，Phenom（羿龙），Athlon（速龙），Sempron（闪龙）等
Lenovo	lenovo	联想，中国领先的 IT 企业。它主要从事台式计算机、笔记本计算机和移动手持设备、服务器和外设的生产、销售。联想的品牌 PC 连续八年一直是中国最畅销的产品。2005 年，它收购了 IBM 全球 PC 业务
Acer	acer	宏碁，中国台湾著名的电脑公司。它主要从事品牌桌上型计算机、笔记本计算机、服务器、液晶显示器及数位家庭（digital home）等产品的研发
HP	hp	惠普，美国著名的打印机、计算机制造商。"HP"为 Hewlett Packard 的缩写，是两位创始人威廉·惠勒（William Hewlett）和大卫·普克（David Packard）姓氏的结合
Sony	SONY	索尼，日本著名数码产品制造商。"SONY"的由来是受到了索尼创立人之一盛田昭夫最喜欢的歌"阳光男孩"（Sunny Boy）影响，起初为 Sonny，意为"活泼可爱的小宝贝"
Dell	DELL	戴尔，Dell 公司是创始人和首席执行官 Michael Dell 的智慧结晶，现已成为世界上第一号 PC 直销商。通过资源最丰富的 Internet 计算机超级市场，Dell 的"用户自配置计算机"Web 网点每天的销售额高达 1 百万美元

续表

Name	Logo	Introduction
Philips	PHILIPS sense and simplicity	飞利浦，荷兰飞利浦公司，主要生产彩显、光驱、家用电器等
Samsung	SAMSUNG	三星，韩国三星公司，著名的彩显制造商，也生产光驱、家用电器等
Seagate	Seagate	希捷，著名的硬盘制造商。希捷为满足今天的消费者及明天的应用需要提供了先进的数字存储解决方案
Epson	EPSON EXCEED YOUR VISION	爱普生，日本爱普生公司，著名的打印机制造商，扫描仪、计算机、投影机等电子设备也是该公司的主要产品。"EPSON"一词源自1964年东京奥运会期间使用的EP-101型打印机，意为Electric Printer的儿子们（SON），EP-101可以算是爱普生公司成长的起点

Part II Hardware Brands

电子产品排行榜如表 3-2～表 3-4 所示。

表 3-2　电子产品排行榜 1

Rank	Mainboard	CPU	Computer memory	HDD
1	GIGABYTE 技嘉	intel 超越未来 Intel	Kingston TECHNOLOGY 金士顿	Seagate 希捷
2	ASUS 华硕	AMD AMD	ADATA 威刚 威刚	WD Western Digital 西部数据
3	msi 微星科技 微星	Cyrix Cyrix	Apacer 宇瞻 宇瞻	HITACHI Inspire the Next 日立
4	BIOSTAR 映泰		CORSAIR 海盗船	SAMSUNG 三星
5	COLORFUL 七彩虹 七彩虹		GeIL 金邦科技	TOSHIBA 东芝

CHAPTER 03　Famous Computer Companies（IT业内的知名公司）

表 3-3　电子产品排行榜 2

Rank	Graphics adapter	Sound card	NIC	CD-ROM
1	七彩虹	创新	Intel	先锋
2	影驰	德国坦克	D-Link	LG
3	迪兰恒进	乐之邦	TP-LINK	三星
4	双敏	华硕	腾达	索尼
5	微星	B-Link	B-Link	华硕

表 3-4　电子产品排行榜 3

Rank	LCD	Mouse	Keyboard	Loudspeaker
1	三星	罗技	罗技	漫步者
2	AOC	雷蛇	雷柏	麦博
3	长城	双飞燕	雷蛇	惠威
4	LG	雷柏	戴尔	三诺

续表

Rank	LCD	Mouse	Keyboard	Loudspeaker
5	PHILIPS 飞利浦	Microsoft 微软	双飞燕 我们一起飞	Hussar 轻骑兵 轻骑兵

（注：以上数据源自 http://www.zol.com.cn/网站，即"中关村在线"，时间为 2010 年 1 月）

Exercises

1. Blank.

（1）世界上最大两家的 CPU 制造商是_____和_____。

（2）世界上最著名的 PC 直销商是_____。

（3）图形图像处理软件 Photoshop 是_____公司的产品。

（4）PC 包括_____和_____。

（5）PC 使用的操作系统绝大多数是_____公司生产的。

2. Thinking.

（1）你知道 Adobe 的流行产品 Adobe Photoshop，Adobe Dreamweaver，Adobe Flash，Adobe Premiere，Adobe Illustrator，Adobe InDesign，Adobe After Effects，Adobe Reader 的主要功能吗？

（2）著名导演詹姆斯·卡梅隆执导的电影《阿凡达》（*Avatar*）是一部科幻电影，你了解该影片使用了哪些 Adobe 设计软件吗？

3. Understand the meaning of the following abbreviations and its full name in English.

DIY
PDF
CPU
HDD
LCD
PDF

参 考 译 文

随着微型计算机的发展，IT 企业获得了一个发展的好时机。如果要配置一台计算机，你必须充分了解各种计算机硬件的制造商，知道该选择什么显示器、CPU、主板等；如果要安装自己需要的软件来完成学习工作或娱乐，你就必须比较各软件商所开发的软件。现在我们就来了解一下一些著名的 IT 企业及计算机硬件品牌。

第二篇

硬件篇

CHAPTER

04

Computer Knowledge
（计算机基础）

There are many types of microcomputers. Take a look at their main hardware components: the system unit, the input devices and the output devices. The system unit consists of the CPU, which is the core of the computer, and the memory. There are two kinds of memories: ROM and RAM. The most common input devices include the keyboard, mouse and scanner. Output devices include the monitor and printer.

Lesson 1　Commonly Used Hardware
（常见的硬件设备）

Interrelated Knowledge

　　计算机系统包括硬件系统和软件系统两大部分。硬件是指组成计算机的各种物理设备，简单地说就是指那些看得见，摸得着的实际物理设备。它包括计算机的主机和外部设备。主机箱里面有主板、CPU、内存、电源、硬盘驱动器、光驱和插在主板总线扩展槽上的各种系统功能扩展卡等。外部设备分为输入设备和输出设备，常见的输入设备有键盘、鼠标、扫描仪，输出设备有显示器、打印机、音响等。

　　计算机能为你做些什么？

　　上网进行网际信息交流获取网络信息服务；

　　购物方式的转变；

　　提供全球的教育资源；

　　提供更为便捷的沟通方式；

　　丰富你的娱乐生活。

Words and Expressions

component	[kəm'pəunənt]	n. 元件，组成部分，成分　　adj. 组成的，构成的
consist	[kən'sist]	vi. 由……组成，构成，在于，符合
core	[kɔ:]	n. 核心，要点
common	['kɔmən]	adj. 常见的，普遍的，共同的
include	[in'klu:d]	vt. 包括，包含
reset	['ri:'set]	v. 重新设定，重新放置，将……恢复原位

Part Ⅰ　PC

桌面计算机如图 4-1 所示。

图 4-1　桌面计算机

Part Ⅱ Keyboard

The keyboard is used to type information into the computer. The standard keyboard has 101 keys.

键盘如图 4-2 所示。

图 4-2 键盘

键盘上的功能键如表 4-1 所示。

表 4-1 键盘上的英语——功能键

Key Name	Function	Key Name	Function
Esc (escape)	退出	Scroll Lock	滚动锁定
Tab (tabulator)	制表	Pause	暂停
Caps Lock	大写锁定	Break	中断
Shift	上挡	Insert	插入
Ctrl (control)	控制	Delete	删除
Alt (alternative)	换挡	Home	起始
Backspace	退格	End	结束
Enter	回车	PgUp (Page Up)	上页
Prtsc (print screen)	打印屏幕	PgDn (Page Down)	下页
SysRq (system request)	系统请求	Num Lock	数字锁定

键盘上的快捷键如表 4-2 所示。

表 4-2 键盘上的英语——常用的快捷键

keyboard shortcuts	Function	Related Word	Remark
Ctrl+A	全选	all	由于大多数人的习惯是用右手操作鼠标,所以将常用的快捷键安排在了方便左手操作的位置,以便于左右手的协作。Z、X、C、V 和 A、S 这几个字母的键位都在最左边,且离 Ctrl 键较近。定义快捷键时,会优先考虑其词义与首字母的吻合,如 all、save、copy 等,以便于记忆,而后考虑其他因素
Ctrl+S	保存	save	
Ctrl+Z	撤消		
Ctrl+X	剪切		
Ctrl+C	复制	copy	
Ctrl+V	粘贴		

Part Ⅲ Commonly Used Hardware

常用的计算机硬件设备如表 4-3 所示。

表 4-3 常用的计算机硬件设备

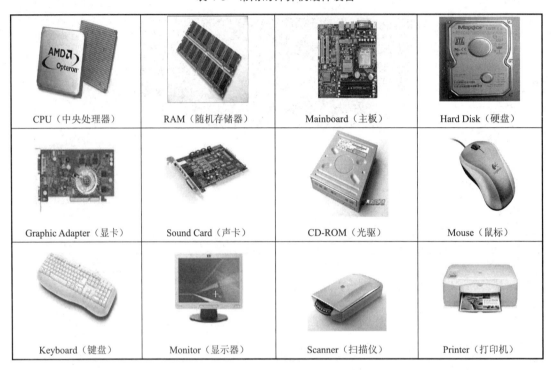

Lesson 2 Booting Computer
（计算机的启动）

❋━ Interrelated Knowledge

在计算机的启动过程中有一个非常完善的硬件自检机制，它会在计算机上电自检那短暂的几秒内完成。其启动过程是在主板 BIOS 的控制下进行的，我们也常把它称为"系统 BIOS"。无论是冷启动还是热启动，系统 BIOS 都会不停地重复硬件检测和引导过程，正是这个不起眼的过程保证了我们可以正常启动和使用计算机。

启动系统时通常有三种方式：冷启动、热启动和复位启动。冷启动的过程包括上电、全面自检、系统引导及初始化等；热启动和冷启动的区别是不需要重新上电、自检的范围很小。在 Windows 环境下，出现死机时，首先应该同时按下"Ctrl＋Alt＋Del"三个键，退出当前的故障程序，而不必重新启动系统。如果无效，则可以再次按下上述三键进行热启动。但是在有些情况下，键盘已经被封锁，系统不能响应键盘输入，这时就必须复位启动了。复位按键一般在机箱上。但是国外的一些品牌机没有安装复位按键,那么就只能冷启动了。由于冷启动过程包含了重新上电的过程,上电过程中的浪涌电流较大,所以反复地冷启动对计算机电源有一定影响。一般对于死机后重启的方选择式而言,优先级别顺序应该是热启动、复位启动、冷启动。

Words and Expressions

check	[tʃek]	v. 检查,阻止,核对　　n. 检查,支票,账单
error	['erə]	n. 错误,过失,谬误,误差
series	['siəri:z]	n. 系列,连续,丛书
notify	['nəutifai]	v. 通知,通告,报告
beep	[bi:p]	n. 哔哔声
copyright	['kɔpirait]	n. 版权,著作权　　adj. 版权的

POST: The BIOS software doing its thing.　通电自检:BIOS 执行操作。

计算机的启动过程如图 4-3～图 4-4 所示。

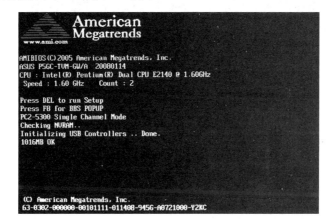

图 4-3　计算机的启动过程 1

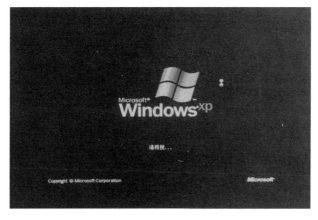

图 4-4　计算机的启动过程 2

如果 BIOS 在通电自检的过程中发现任何错误，它会以一系列的响声提示，或者在屏幕上显示文字信息。在这个阶段出现的错误通常都是硬件问题。

图 4-4 中的 Copyright © Microsoft Corporation 表示版权由微软公司拥有。

计算机启动完成的界面如图 4-5 所示。

图 4-5　计算机启动完成

Exercises

1．Match.
（1）monitor　　A．打印机
（2）printer　　B．扫描仪
（3）mouse　　C．键盘
（4）keyboard　D．显示器
（5）scanner　　E．鼠标

2．Thinking.

（1）图 4-3 的计算机启动过程中屏幕的提示告诉了我们哪些信息？

（2）图 4-6 是一次计算机开机启动时出现的屏幕信息，你知道出现了什么问题吗？

3．Understand the meaning of the following abbreviations and its full name in English.

CPU

ROM

RAM

CRT

LCD

POST

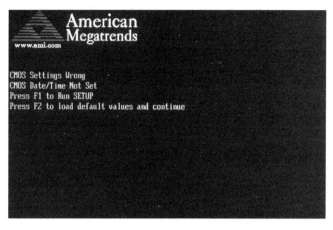

图 4-6　计算机启动过程的屏幕显示

参 考 译 文

微型计算机的种类繁多。我们来了解一下它的硬件的主要组成部分。它包括主机、输入设备和输出设备。主机由中央处理器（CPU）和内存储器组成，其中 CPU 是计算机的核心，内存储器包括 ROM 和 RAM 两种类型；最常见的输入设备有键盘、鼠标和扫描仪，输出设备有显示器和打印机。

CHAPTER 05

BIOS Settings
（BIOS 设置）

New PCs need a few configurations before they can be used. When the computer is started or restarted, we can press "Delete" to enter BIOS SETUP UTILITY after the "Press DEL to run Setup" message comes up on the screen. Generally speaking, different kinds of mainboards have different types of BIOS SETUP UTILITY, but their functions and settings are similar. Let's learn some functions of the BIOS SETUP UTILITY.

Interrelated Knowledge

BIOS 是 Basic Input/Output System（基本输入/输出系统）的缩写，是一组设置硬件的计算机程序，保存在主板上的一块 EPROM 或 EEPROM 芯片中，里面装有系统的重要信息和设置系统参数的设置程序——BIOS Setup 程序。形象地说，BIOS 应该是连接软件程序与硬件设备的一座"桥梁"，负责解决硬件的即时要求。BIOS ROM 芯片不但可以在主板上看到，而且 BIOS 管理功能如何在很大程度上决定了主板性能是否优越。BIOS 的管理功能主要包括以下方面：BIOS 中断服务程序；BIOS 系统设置程序；POST 上电自检；BIOS 系统启动自举程序。

CMOS 即 Complementary Metal Oxide Semiconductor——互补金属氧化物半导体，是主板上的一块可读写的 RAM 芯片，用来保存当前系统的硬件配置和用户对参数的设定，其内容可通过设置程序进行读写。CMOS 芯片由主板上的纽扣电池供电，即使系统断电，其参数也不会丢失。CMOS 芯片只有保存数据的功能，而对 CMOS 中各项参数的修改要通过 BIOS 的设定程序来实现。

由于 BIOS 和 CMOS 都跟系统设置密切相关，很多时候我们会把 BIOS 和 CMOS 混为一谈。所以在实际使用过程中产生了 BIOS 设置和 CMOS 设置两种说法，其实它们指的都是同一回事，但 BIOS 与 CMOS 却是两个完全不同的概念，切勿混淆。完整的说法应该是"通过 BIOS 设置程序对 CMOS 参数进行设置"。平常所说的 CMOS 设置和 BIOS 设置是其简化说法，这也就在一定程度上造成了两个概念的混淆。

事实上，BIOS 与 CMOS 既相关又不同：BIOS 中的系统设置程序是完成 CMOS 参数设置的手段；CMOS RAM 既是 BIOS 设定系统参数的存放场所，又是 BIOS 设定系统参数的结果。

什么时候需要对 BIOS 或 CMOS 进行设置呢？

众所周知，BIOS 或 CMOS 设置是由操作人员根据计算机实际情况而人工完成的一项十分重要的系统初始化工作。在以下情况下，必须进行 BIOS 或 CMOS 设置：

（1）新购计算机；
（2）新增设备；
（3）CMOS 数据意外丢失；
（4）系统优化。

Words and Expressions

setup	['setʌp]	n. 设置，安装，计划
configure	[kən'figə]	v. 配置
configuration	[kən,figju'reiʃən]	n. 结构，形状，【计算机】配置
disable	[dis'eibl]	v. 使……失去能力，【计算机】禁用
enable	[i'neibl]	vt. 使……能够，【计算机】启用，激活
supervisor	['sju:pəvaizə]	n. 监督人，管理人，【计算机】（网络）超级用户
function	['fʌŋkʃən]	n. 功能，函数
warning	['wɔ:niŋ]	n. 警告，警报
auto	['ɔ:təu]	pref. 自动的，自己的

screen	[skri:n]	n. 屏，幕，银幕	
similar	['similə]	adj. 类似的，同样的	
press	[pres]	v. 按，压	
system	['sistəm]	n. 系统，体系，制度	
support	[sə'pɔ:t]	n. 支持，援助，供养　vt. 支援，帮助，支持	
priority	[prai'ɔriti]	n. 优先权，优先顺序，优先	
sequence	['si:kwəns]	n. 顺序，次序，序列　vt. 按顺序排好	
normal	['nɔ:məl]	n. 常态，标准　adj. 正常的，正规的	
exit	['ekzit, sit]	v. 退出，离去　n. 出口，退场	
default	[di'fɔ:lt]	n. 假设值，默认（值）　v. 默认，【计算机】默认	

Part Ⅰ　AMI BIOS Software

如图 5-1 所示为 AMI BIOS 的开机画面。目前其主板 BIOS 主要有 Aword BIOS 和 AMI BIOS 两种类型。

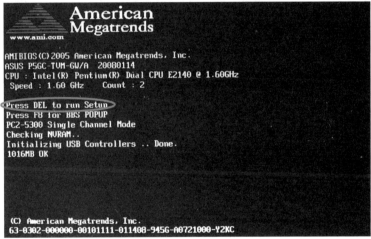

图 5-1　AMI BIOS 的开机画面

➢ Press DEL to run Setup
　按 DEL 键运行 BIOS 设置程序。

Part Ⅱ　Main

基本设置如图 5-2 所示。
帮助提示信息区：

➢ Use [Enter], [Tab] or [Shift-Tab] to select a field.
　使用[Enter] 键，[Tab] 键或[Shift-Tab]组合键选择项目选项。

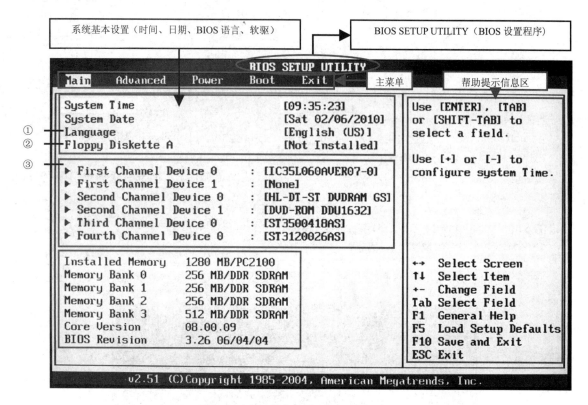

图 5-2　BIOS 功能的设置——基本设置

> Use [+] or [−] to configure system Time.
> 使用[+]键或[−]键设置系统时间。

←→	Select Screen	按←、→键，选择主菜单
↑↓	Select Item	按↑、↓键，选择设置项目
+−	Change Field	按+、−键，改变项目选项
Tab	Select Field	按 Tab 键，选择项目选项
F1	General Help	按 F1 键，获取帮助信息
F5	Load Setup Defaults	按 F5 键，加载默认的设置
F10	Save and Exit	按 F10 键，保存并退出
ESC	Exit	按 ESC 键，退出

> ①<Enter> to select display language for BIOS setup screen.
> 按回车键选择 BIOS 设置界面的语言。

> ②<Enter> to select floppy type.
> 按回车键选择软驱类型。

> ③Select the type of device connected to the system.
> 选择连接到系统的设备类型。

Part Ⅲ　Advanced

高级设置如图 5-3 所示。

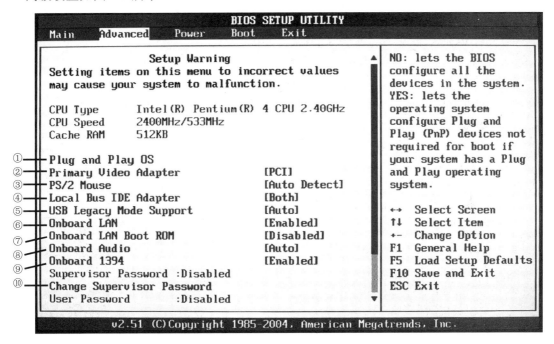

图 5-3　BIOS 功能的设置——高级设置

➢ Setup Warning
　　设置警告。

➢ Setting items on this menu to incorrect values may cause your system to malfunction.
　　对该菜单中的项目进行不恰当的设置可能会导致您的系统故障。

➢ ①No: lets the BIOS configure all the devices in the system.
　　Yes: lets the operating system configure Plug and Play (PnP) devices not required for boot if your system has a Plug and Play operating system.
　　No：由 BIOS 配置计算机系统的所有设备。
　　Yes：如果操作系统支持即插即用功能，则由操作系统配置不需启动的即插即用设备。

➢ ②Select which graphics controller to use as the primary boot device.
　　选择用做主引导设备的图形控制器。

➢ ③Select support for PS/2 Mouse.
　　选择支持 PS/2 鼠标。

- ④Disabled: disables the integrated IDE Controller.
  Primary: enables only the Primary IDE Controller.
  Secondary: enables only the Secondary IDE Controller.
  Both: enables both IDE Controllers.
  Disabled：禁用集成 IDE 控制器。
  Primary：仅启用主 IDE 控制器。
  Secondary：仅启用从 IDE 控制器。
  Both：同时启用主从 IDE 控制器。

- ⑤Enables support for legacy USB. AUTO option disables legacy support if no USB devices are connected.
  启用支持传统 USB。如果 USB 设备没有连接，AUTO 选项将不能支持传统 USB。

- ⑥Enable/Disable Onboard Lan.
  启用/禁用主板集成网卡。

- ⑦Onboard LAN boot ROM Configuration.
  主板集成网卡启动 ROM 配置。

- ⑧Enable/Disable AC97 Audio Controller function.
  启用/禁用 AC97 音频控制器功能。

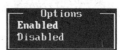

- ⑨Enable/Disable Onboard1394。
  启用/禁用主板集成 1394 接口。

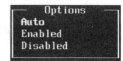

- ⑩<Enter> to change password.
  <Enter> again to disable password.
  按<Enter>键更改密码。
  再次按<Enter>键禁用密码。

Part Ⅳ Power

电源管理设置如图 5-4 所示。

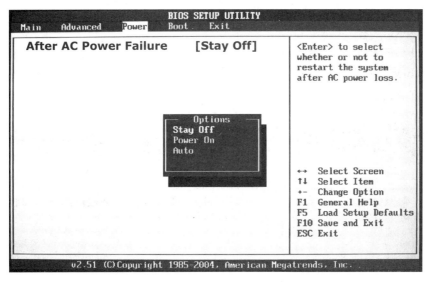

图 5-4　BIOS 功能的设置——电源管理设置

➢ <Enter> to select whether or not to restart the system after AC power loss.
按<Enter>键选择断电后恢复供电时是否重新启动系统。

Part Ⅴ　Boot

启动设置如图 5-5 所示。

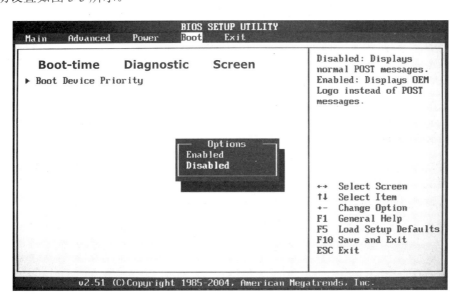

图 5-5　BIOS 功能的设置——启动设置 1

➢ Disabled: Displays normal POST messages.
  Disabled：显示正常的自检信息。
➢ Enabled: Displays OEM Logo instead of POST messages.
➢ Enabled：显示 OEM 标志而不是自检信息。

启动设置 2 如图 5-6 所示。

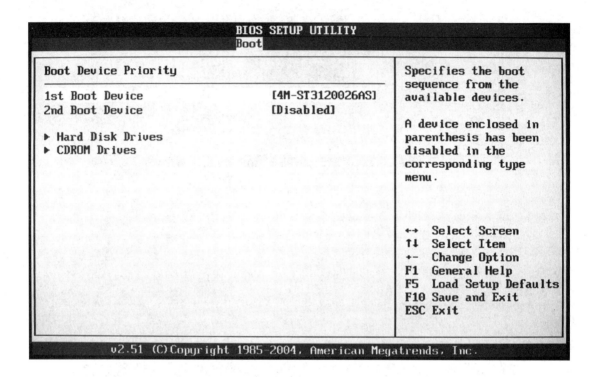

图 5-6 BIOS 功能的设置——启动设置 2

➢ Specifies the boot sequence from the available devices.
  指定现有设备的启动顺序。
➢ A device enclosed in parenthesis has been disabled in the corresponding type menu.
  在相应类型的菜单中括号里的设备将不显示。
➢ Specifies the Boot Device Priority sequence from available CDROM Devices.
  指定现有光驱设备的优先启动顺序。
➢ Specifies the Boot Device Priority sequence from available Hard Devices.
  指定现有硬盘设备的优先启动顺序。

Part Ⅵ Exit

退出设置如图 5-7 所示。

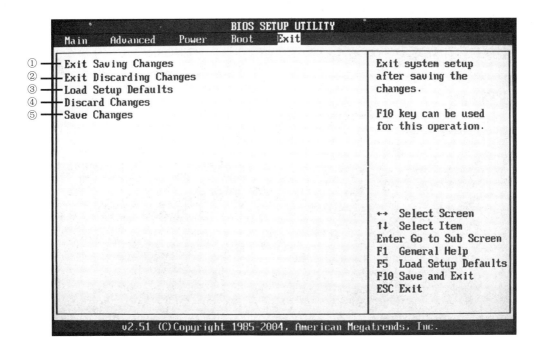

图 5-7 BIOS 功能的设置——退出设置

> ①Exit system setup after saving the changes.
> F10 key can be used for this operation.
> 保存更改并退出系统。
> 按 F10 键进行此项操作。

> ②Exit system setup without saving any changes.
> ESC key can be used for this operation.
> 放弃更改并退出系统。
> 按 ESC 键进行此项操作。

> ③Load Optimal Default values for all the setup questions.
> F5 key can be used for this operation.
> 为所有的设置加载最佳默认值。
> 按 F5 键进行此项操作。

> ④Discards changes done so far to any of the setup questions.
> F7 key can be used for this operation.
> 放弃迄今为止的任何更改设置。
> 按 F7 键进行此项操作。

> ⑤Save changes to COMS.
> 保存更改到 CMOS。

Exercises

1. Thinking.

请说出在图 5-8 中给 AWARD BIOS 设置开机密码（管理员密码）、电源管理、保存退出时分别使用哪一项，以及图下方各功能键的作用。

```
              CMOS SETUP UTILITY
             AWARD SOFTWARE, INC.

STANDARD CMOS SETUP              INTEGRATED PERIPHERALS
BIOS FEATURES SETUP              SUPERVISOR PASSWORD
CHIPSET FEATURES SETUP           USER PASSWORD
POWER MANAGEMENT SETUP           IDE HDD AUTO DETECTION
PNP/PCI CONFIGURATION            SAVE & EXIT SETUP
LOAD BIOS DEFAULTS               EXIT WITHOUT SAVING
LOAD OPTIMUM SETTINGS

Esc : Quit                      ↑↓→← : Select Item
F10 : Save & Exit Setup         <Shift>F2 : Change Color
```

图 5-8 CMOS 功能设置的主菜单

2. Understand the meaning of the following abbreviations and its full name in English.

BIOS
CMOS
PnP

参 考 译 文

一台新的计算机需要进行一些设置才能使用。当开启计算机或重新启动计算机后，屏幕显示"按 DEL 键运行设置程序"信息时，按下 Delete 键即可进入 BIOS 的设置程序。一般不同的主板的 BIOS 是有区别的，但是其功能和设置大同小异。让我们一起来了解一下 BIOS 的功能吧！

CHAPTER | 06

Installing and Uninstalling
（安装与卸载）

A PC contains a hardware system and a software system. In order for the computer work efficiently, we must install the operating system and necessary applications. The operating system is the main software that controls and runs the operations of the computer hardware. It also provides the interface through which the user communicates with the computer. Various types of application software such as office, communications, multimedia software, etc., are used for certain jobs or tasks. When software is no longer used, you can uninstall it to obtain more available space.

Lesson 1　Installing Operating System
（系统安装）

❀— Interrelated Knowledge

计算机系统由硬件系统和软件系统组成。没有软件的计算机称为"裸机"，什么事情也做不了。计算机必须在软件的支持下才能工作，因此软件的安装是计算机实现高效工作的重要保障。计算机软件可分为系统软件和应用软件两大类。

操作系统是控制其他程序运行，管理系统资源并为用户提供操作界面的系统软件的集合。计算机操作系统的类型很多，目前一般的个人计算机中最常用的操作系统是 Microsoft Windows，其次是苹果公司的 Mac 操作系统。

Windows XP 的安装过程基本不需要人工干预，但是有些地方、项目还是需要我们关注的。如在系统安装前，首先需在 BIOS 中将光驱设置为第一启动项和选择系统安装分区；在系统安装时，需完成序列号的输入和网络连接的设置；而在系统安装后，需进行用户账号的创建等。

Ghost 是美国赛门铁克公司（Symantec）推出的一款出色的硬盘备份还原工具，俗称克隆软件，主要用于系统、数据的备份与恢复。使用 Ghost 软件可迅速方便地实现系统的快速安装和恢复。它的工作原理和方法不同于其他备份软件，它是将硬盘的一个分区或整个硬盘作为一个对象来操作，可以完整复制对象，并将其打包压缩成为一个镜像文件（GHO），在需要的时候，又可以通过它本身将该镜像文件快速恢复到指定的分区或对应的硬盘中。其功能包括两个硬盘之间的对拷、两个硬盘的分区对拷、两台计算机之间的硬盘对拷、制作硬盘的镜像文件等。用得最多的是它的分区备份和恢复功能，它能够先将硬盘的一个分区压缩备份成镜像文件，然后存储在另一个分区硬盘或大容量存储器中，万一原来的分区发生问题，就可以将所备份的镜像文件拷回去，让分区恢复正常了。

❀— Words and Expressions

install	[in'stɔ:l]	vt. 安装，安置
contain	[kən'tein]	vt. 包含，容纳
efficiently	[i'fiʃəntli]	adv. 有效地
communicate	[kə'mju:nikeit]	v. 交流，传达，沟通
multimedia	[mʌlti'mi:diə]	adj. 多媒体的　n. 多媒体
inspect	[in'spekt]	vt. 调查，检阅
image	['imidʒ]	n. 图像，影像　vt. 想象，描绘，反映
partition	[pɑ:'tiʃən]	vt. 区分，隔开，分割　n. 分割，隔离物
compress	[kəm'pres]	vt. 压缩，压榨
statistics	[stə'tistiks]	n. 统计，统计数字，统计学
percent	[pə'sent]	n. 百分比　adv. 百分之……　adj. 百分之……的
elapse	[i'læps]	v. 逝去，过去

Part Ⅰ　Installing Windows XP

首先在 BIOS 中将光驱设置为第一启动项，如图 6-1 所示。

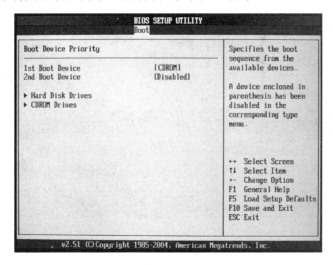

图 6-1　在 BIOS 中将光驱设置为第一启动项

Windows XP 的安装过程具体如图 6-2～图 6-6 所示。

图 6-2　Windows XP 的安装过程 1　　　图 6-3　Windows XP 的安装过程 2

➢ Press any key to boot from CD…
按任意键即由光盘启动。
➢ Setup is inspecting your computer's hardware configuration…
安装程序正在检查计算机的硬件配置。
➢ Press F6 if you need to install a third party SCSI or RAID driver…
如果需要安装第三方 SCSI 或 RAID 驱动按 F6 键。

图 6-4　Windows XP 的安装过程 3

➢ Setup is loading files (Windows Executive)…
  安装程序正在调用文件（Windows 执行）。

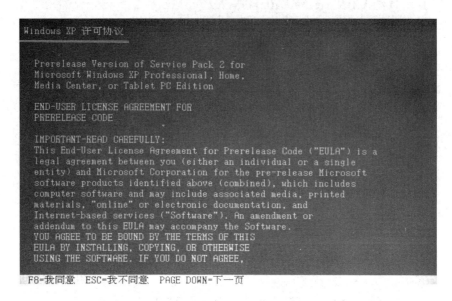

图 6-5　Windows XP 的安装过程 4

图 6-6　Windows XP 的安装过程 5

Part Ⅱ　Ghost

1. To Image

以分区备份（To Image）为例，了解使用 Ghost 进行系统备份的过程。硬盘的复制（Local—Disk——To Disk）和硬盘备份（Local—Disk—To Image）操作与分区备份相似。
Operating interface of 'To Image' for Symantec Ghost 11.0.2（如图 6-7～图 6-15 所示）。

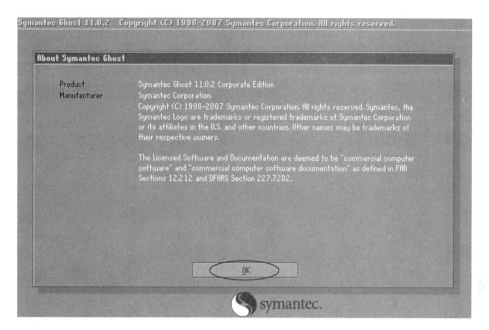

图 6-7　Symantec Ghost 的主界面

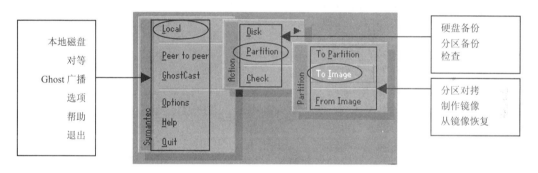

图 6-8　Ghost 主菜单之"制作镜像"

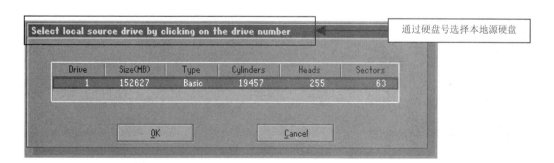

图 6-9　选择本地源硬盘

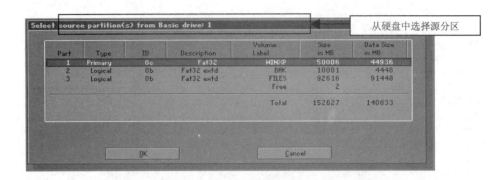

图 6-10 选择源分区

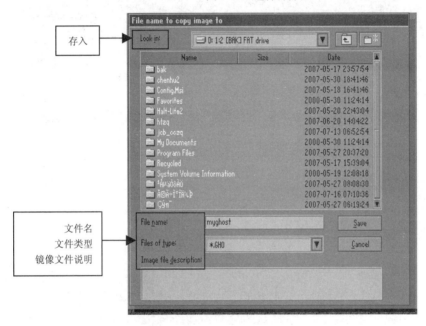

图 6-11 选择存放镜像文件的路径并输入镜像文件名

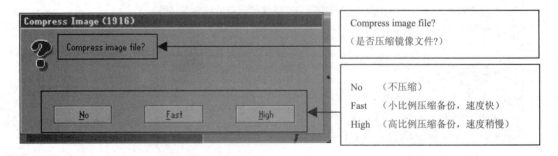

图 6-12 压缩镜像文件

CHAPTER 06　Installing and Uninstalling（安装与卸载）

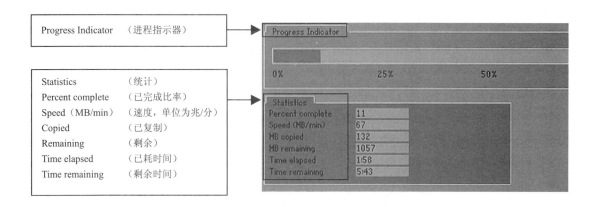

图 6-13　进行分区备份

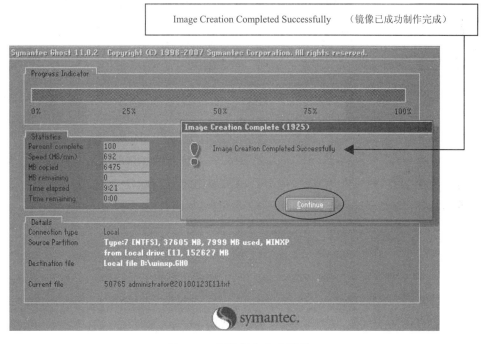

图 6-14　镜像制作完成画面

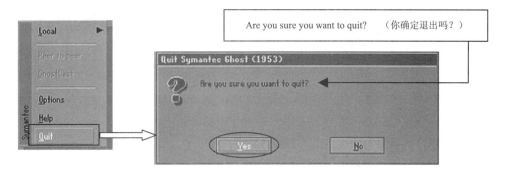

图 6-15　退出操作

2. From Image

备份还原，以恢复备份的分区为例（From Image），其他还原操作类似。

Operating interface of 'From Image' for Symantec Ghost 11.0.2　（如图 6-16～图 6-23 所示）。

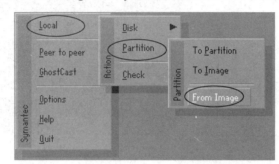

图 6-16　Ghost 主菜单之"从镜像恢复"

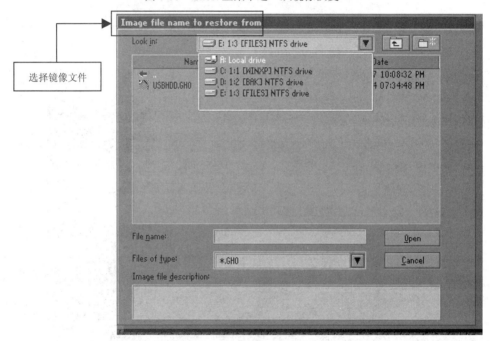

图 6-17　选择镜像文件

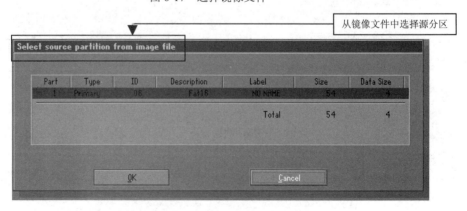

图 6-18　选择镜像文件中的源分区

CHAPTER 06　Installing and Uninstalling（安装与卸载）

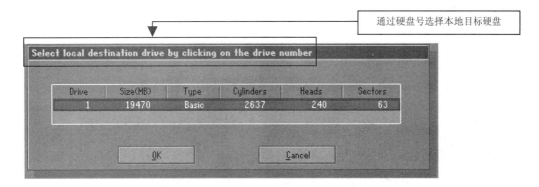

图 6-19　选择本地目标硬盘

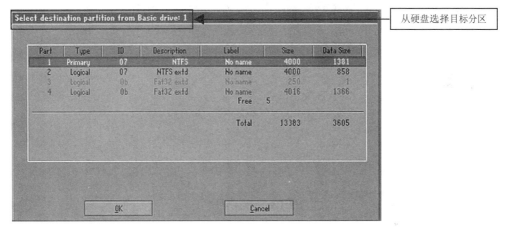

图 6-20　选择目标分区

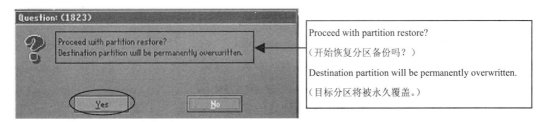

图 6-21　是否恢复备份

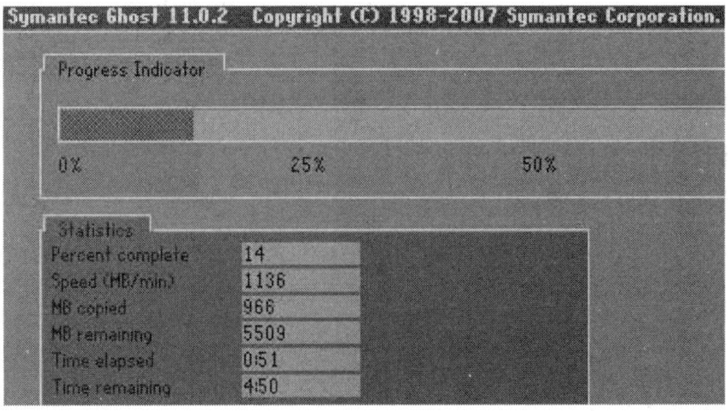

图 6-22　正在恢复备份

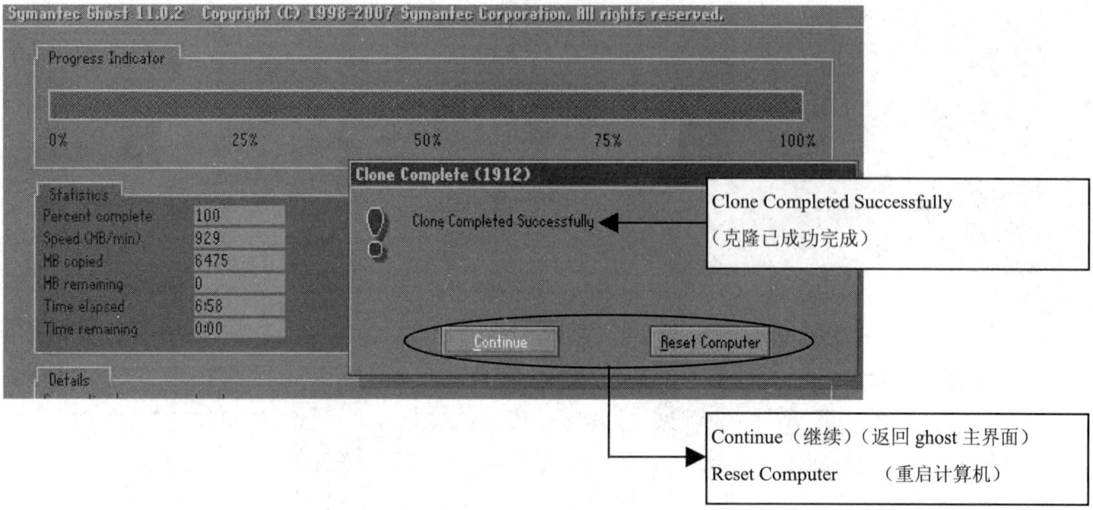

图 6-23　备份完成，重启计算机使设置生效

Lesson 2　Installing　Device Drivers
（安装设备驱动程序）

❀ Interrelated Knowledge

　　设备驱动程序的英文名为"Device Driver"。它是一种可以使计算机和设备通信的特殊程序，即相当于硬件的接口。操作系统只有通过这个接口，才能控制硬件设备的工作，假如某设备的驱动程序未能正确安装，便不能正常工作。因此，设备驱动程序有"硬件的灵魂"之称，是硬件和系统之间的桥梁。

　　驱动程序在系统中所占的地位十分重要，一般当操作系统安装完毕后，首要的便是安装硬件设备的驱动程序。从理论上讲，所有的硬件设备都需要安装相应的驱动程序才能正常工作。不过在大多数情况下，我们并不需要安装所有硬件设备的驱动程序，如硬盘、显示器、光驱、键盘、鼠标等就不需要安装驱动程序，而显卡、声卡、扫描仪、摄像头、Modem 等则需要安装驱动程序。这主要是由于前者对于一台个人计算机来说是必需的，所以早期的设计人员将这些硬件列为 BIOS 能直接支持的硬件。换句话说，上述硬件安装后就可以被 BIOS 和操作系统直接支持，不再需要安装驱动程序了。另外，不同版本的操作系统对硬件设备的支持也是不同的，一般情况下版本越高所支持的硬件设备也越多，因此也存在装好系统后一个驱动程序也不用安装便能正常使用的情况。

CHAPTER 06　Installing and Uninstalling（安装与卸载）

Words and Expressions

wizard　　　　　　['wizəd]　　　　　n. 向导；有特殊才干的人，奇才；术士
Ethernet　　　　　['i:θənet]　　　　 n. 以太网

Windows 下的设备驱动程序的安装非常简便，下面以网卡的驱动程序安装为例。
Installing NIC Driver（如图 6-24～图 6-27 所示）。

图 6-24　在设备驱动程序文件夹中选择打开网卡——LAN 的驱动程序文件夹

图 6-25　双击安装程序文件——SETUP.EXE 进行网卡的安装

图 6-26　网卡驱动程序的安装过程

➢ Now waiting for finding devices and installing
　等候查找设备并安装（驱动程序）。

图 6-27　完成网卡驱动程序的安装

➢ REALTAK Gigabit and Fast Ethernet NIC Driver
　REALTAK 千兆快速以太网网卡驱动程序。
　注：REALTAK（瑞昱）——网卡生产厂商
➢ Installshield Wizard
　安装向导。

Lesson 3 Installing Applications
（安装应用软件）

Interrelated Knowledge

要发挥计算机的最大效能，更好地为我们的工作、学习和娱乐服务，仅有操作系统还远远不够，还需要安装更多的软件。不同的软件具有不同的作用，如 Microsoft Office 是常用办公软件，ACDSee 是图像浏览工具，金山毒霸具有查杀病毒的功能，WinRAR 是压缩工具，迅雷是下载工具，Adobe Reader 是阅读和打印 PDF 文件的最佳帮手，QQ 可以实现网上聊天等，用户可以根据自己的需要选择安装相应的软件。应用软件的安装方法大体相同，下面将以 Adobe Reader 9.3 的安装为例进行说明。

Adobe Reader（也称 Acrobat Reader）是美国 Adobe 公司开发的一款优秀的 PDF 文档阅读软件。文档的撰写者可以向任何人分发自己制作（通过 Adobe Acrobat 制作）的 PDF 文档而不用担心被恶意篡改。

PDF (Portable Document Format) 文件格式是电子发行文档的事实上的标准，Adobe Acrobat Reader 是一个查看、阅读和打印 PDF 文件的最佳工具。

Words and Expressions

application	[ˌæpli'keiʃən]	n. 应用；应用软件程序
status	['steitəs]	n. 地位，身份，情形，状况
treaty	['tri:ti]	n. 条约，协定

Part Ⅰ Download Adobe Reader 9.3

下载 Adobe Reader 9.3，如图 6-28 所示。

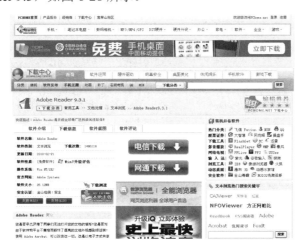

图 6-28 下载 Adobe Reader 9.3

Part Ⅱ Install Adobe Reader 9.3

安装 Adobe Reader 9.3，如图 6-29～图 6-34 所示。

图 6-29 安装 Adobe Reader 9.3 的过程 1

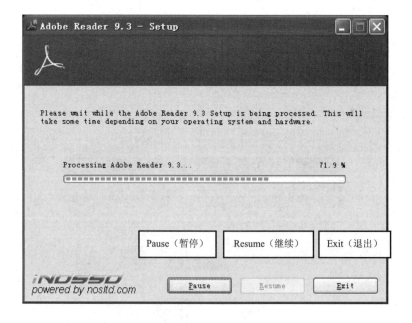

图 6-30 安装 Adobe Reader 9.3 的过程 2

➢ Please wait while the Adobe Reader 9.3 Setup is being processed. This will take some time depending on your operating system and hardware.
Adobe Reader 9.3 安装程序正在处理中,请稍候。根据您的操作系统和硬件配置,这可能需要一些时间。

➢ Processing Adobe Reader 9.3 . . .
正在处理 Adobe Reader 9.3……

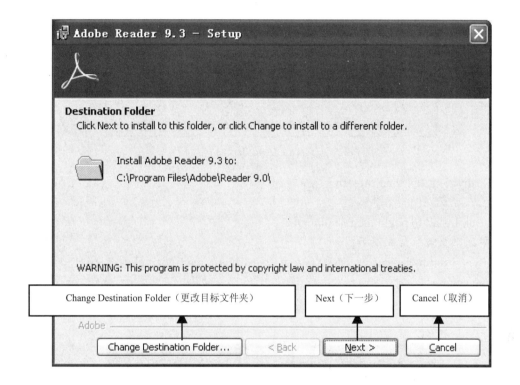

图 6-31　安装 Adobe Reader 9.3 的过程 3

➢ Destination Folder
目的地文件夹。

➢ Click Next to install to this folder, or click Change to install to a different folder.
单击"下一步"按钮安装到此文件夹,或单击"更改"安装到不同的文件夹。

➢ Install Adobe Reader 9.3 to :
安装 Adobe Reader 9.3 至:

➢ WARNING: This program is protected by copyright law and international treaties.
警告:本程序受版权法和国际条约的保护。

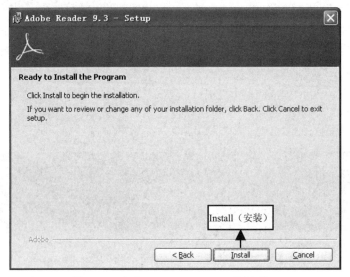

图 6-32　安装 Adobe Reader 9.3 的过程 4

➢ Ready to Install the Program
已做好安装程序的准备。

➢ Click Install to begin the installation.
单击"安装"开始安装。

➢ If you want to review or change any of your installation folder, click Back. Click Cancel to exit setup.
如果您需要检查或更改安装文件夹，请单击"上一步"按钮。单击"取消"按钮可退出安装。

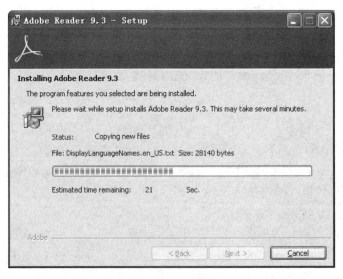

图 6-33　安装 Adobe Reader 9.3 的过程 5

➢ Installing Adobe Reader 9.3
正在安装 Adobe Reader 9.3。

➢ The program features you selected are being installed.
正在安装您选择的程序功能。
➢ Please wait while setup installs Adobe Reader 9.3. This may take several minutes.
请等待安装程序安装 Adobe Reader 9.3。此操作可能花费几分钟时间。
➢ Status: Copying new files
状态：正在复制新文件。
➢ Estimated time remaining: 21 Sec.
估计剩余时间：21 秒。

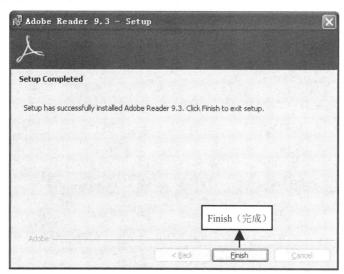

图 6-34　安装 Adobe Reader 9.3 的过程 6

➢ Setup Completed
安装完成。
➢ Setup has successfully installed Adobe Reader 9.3. Click Finish to exit setup.
安装程序已经成功地安装了 Adobe Reader 9.3。请单击"完成"按钮来退出安装程序。

Lesson 4　Uninstalling Applications（卸载应用程序）

Interrelated Knowledge

对于一些已经存在的不使用的程序，可以通过卸载的方法让系统将其删除，从而释放磁盘空间。在 Windows XP 中删除应用程序通常有如下的两种方法。

（1）通过控制面板中的"添加或删除程序"卸载应用程序。

（2）使用反安装程序卸载应用程序。其中常见的有两种形式：一是单击"开始"按钮，在"所有程序"中选择需删除的应用程序，单击其下自带的"卸载……"选项；二是部分应用程序可通过本身的安装程序自带的"程序维护"功能来删除。

Words and Expressions

panel	['pænl]	n. 面板，嵌板，仪表盘　v. 嵌镶
program	['prəugræm]	n. 程序，计划　vt. 编制程序，拟……计划
maintenance	['meintinəns]	n. 维护，保持，维修
remove	[ri'mu:v]	v. 消除，除去，脱掉，搬迁
corrupt	[kə'rʌpt]	adj. 有缺陷的；有错误的　v. 引起……错误；破坏
shortcut	['ʃɔ:tkʌt]	n. 捷径；快捷方式，快捷键
registry	['redʒistri]	n. 注册，登记，登记簿
entry	['entri]	n. 进入，入口，登记，条目

Part Ⅰ　Using control panel

利用控制面板卸载应用程序的步骤如图 6-35～图 6-39 所示。

图 6-35　步骤一：打开"控制面板"

图 6-36　步骤二：选择"添加或删除程序"

图 6-37　步骤三：选择列表中需删除的应用程序，单击"删除"按钮

图 6-38　步骤四：单击弹出的对话框中
　　　　的"是"按钮，确认删除操作

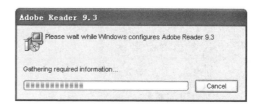

图 6-39　步骤五：等待完成删除过程

- Please wait while Windows configures Adobe Reader 9.3
  请等待 Windows 配置 Adobe Reader 9.3。
- Gathering required information…
  正在收集所需信息……

Part II　Using Uninstalling program

（1）通过"开始"菜单删除应用程序，如图 6-40 所示。

图 6-40　卸载 ICQ7

（2）通过安装程序的"程序维护"功能删除程序，如图 6-41～图 4-47 所示。

图 6-41　删除 Adobe Reader 9.3 的步骤 1

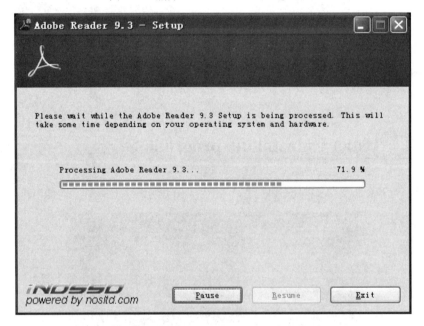

图 6-42　删除 Adobe Reader 9.3 的步骤 2

➢ Please wait while the Adobe Reader 9.3 Setup is being processed. This will take some time depending on your operating system and hardware.
Adobe Reader 9.3 安装程序正在处理中，请稍候。根据您的操作系统和硬件配置，这可能需要一些时间。

➢ Processing Adobe Reader 9.3 …
  正在处理 Adobe Reader 9.3 …

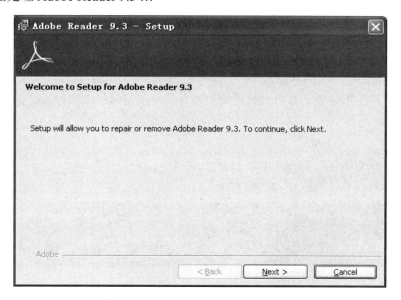

图 6-43　删除 Adobe Reader 9.3 的步骤 3

➢ Welcome to Setup for Adobe Reader 9.3
  欢迎使用 Adobe Reader 9.3 安装程序。

➢ Setup will allow you to repair or remove Adobe Reader 9.3. To continue, click Next.
  安装程序将允许您修复或删除 Adobe Reader 9.3。要继续，请单击"下一步"按钮。

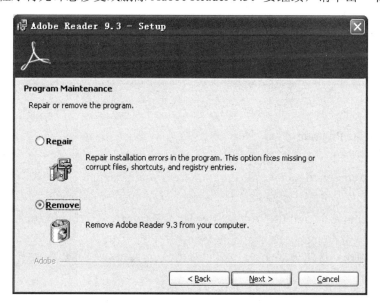

图 6-44　删除 Adobe Reader 9.3 的步骤 4

➢ Program Maintenance
  程序维护。

- Repair or remove the program.
  修复或删除程序。
- Repair
  修复。
- Repair installation errors in the program. This option fixes missing or corrupt files, shortcuts, and registry entries.
  修复程序中的错误。通过此选项您可修复缺少或损坏的文件、快捷方式和注册表项。
- Remove
  删除。
- Remove Adobe Reader 9.3 from your computer.
  从您的计算机上删除 Adobe Reader 9.3。

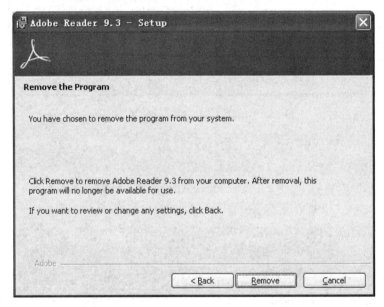

图 6-45　删除 Adobe Reader 9.3 的步骤 5

- Remove the Program
  删除程序。
- You have chosen to remove the program from your system.
  您已经选择从系统中删除此程序。
- Click Remove to remove Adobe Reader 9.3 from your computer. After removal, this program will no longer be available for use.
  单击"删除"按钮将从您的计算机中删除 Adobe Reader 9.3。删除之后，本程序将不再可用。
- If you want to review or change any settings, click Back.
  要查看或更改任何设置，请单击"上一步"按钮。

CHAPTER 06   Installing and Uninstalling（安装与卸载）

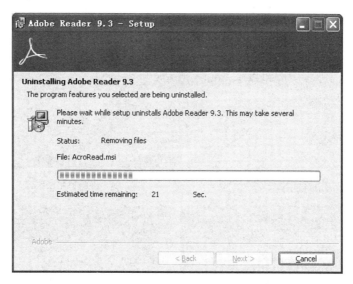

图 6-46　删除 Adobe Reader 9.3 的步骤 6

- ➤ Uninstalling Adobe Reader 9.3
  卸载 Adobe Reader 9.3。

- ➤ The program features you selected are being uninstalled.
  正在卸载您选择的程序功能。

- ➤ Please wait while setup uninstalls Adobe Reader 9.3. This may take several minutes.
  请等待安装程序卸载 Adobe Reader 9.3。此操作可能花费几分钟时间。

- ➤ Status: Removing files
  状态：正在删除文件。

- ➤ Estimated time remaining: 21 Sec.
  估计剩余时间：21 秒。

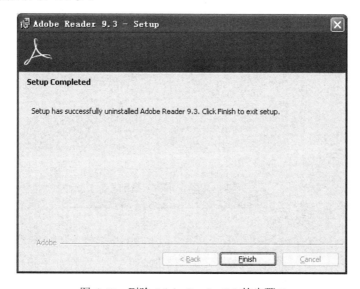

图 6-47　删除 Adobe Reader 9.3 的步骤 7

➢ Setup Completed
安装完成。
➢ Setup has successfully uninstalled Adobe Reader 9.3. Click Finish to exit setup.
安装程序已经成功地卸载了 Adobe Reader 9.3。请单击"完成"按钮来退出安装程序。

Exercises

1. Match (According to Fig. 6-48).

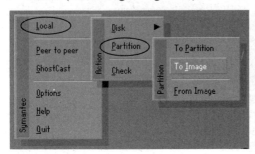 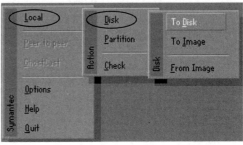

图 6-48 Ghost 菜单

（1）partition to partition   A. 从镜像还原到分区
（2）disk to disk   B. 本地磁盘
（3）ghost   C. 硬盘备份
（4）partition to image   D. 备份分区
（5）check   E. 硬盘复制
（6）options   F. 备份
（7）disk to image   G. 分区复制
（8）partition from image   H. 选项
（9）local   I. 检查（硬盘或备份的文件）

2. Understand the meaning of the following abbreviations and its full name in English.
LAN
VGA
NIC

参 考 译 文

一台计算机包括硬件系统和软件系统。我们必须为计算机安装操作系统和必需的应用软件，只有这样才能实现计算机的高效工作。操作系统是控制和操作计算机硬件的主要软件，它为用户提供操作界面，实现用户和计算机的信息交换。应用软件用来实现特定功能，如办公软件、通信软件、多媒体软件等。当某个软件不再被使用时，你可以卸载它以得到更多可用空间。

第三篇

软件篇

CHAPTER 07

Operating System
（操作系统）

The operating system (OS) is a program that manages computers' hardware and software. It's in charge of managing, adjusting and controlling computer resources, as well as making them work correctly. No computer can operate without the operating system. It plays a role between the resources and the users. During the development of computers, people have created many different operating systems, such as DOS, Mac OS, Windows, Linux, UNIX.[1]

Interrelated Knowledge

操作系统用于管理计算机系统的全部硬件资源（包括软件资源及数据资源）；控制程序运行；改善人机界面；为其他应用软件提供支持等，可使计算机系统所有资源最大限度地发挥作用，为用户提供方便的、有效的、友善的服务界面。

所有的操作系统具有并发性、共享性、虚拟性和不确定性四个基本特征。

操作系统的形态非常多样，不同机器安装的 OS 可从简单到复杂，从手机的嵌入式系统到超级计算机的大型操作系统变化。

许多操作系统制造者对 OS 的定义也不大一致，如有些 OS 集成了图形用户界面，而有些 OS 仅使用文本接口，而将图形界面视为一种非必要的应用程序。

Words and Expressions

operate	['ɔpəreit]	vi. 操作，运转
commercialize	[kə'məːʃəlaiz]	vt. 使商业化
version	['vəːʃən]	n. 版本，说法
christen	['krisn]	vt. 为……命名
interface	['intəfeis]	n. 界面，接口 v. 连接，作接口
workstation	['wəːk.steiʃən]	n. 工作站
hint	[hint]	n. 暗示
tip	[tip]	n. 提示
diagnosis	[daiəg'nəusis]	n. 诊断
switch	[switʃ]	v. 转换，改变，交换 n.【计】转换器
compare	[kəm'pɛə]	v. 比较，比喻，对照
parameter	[pə'ræmitə]	n. 参数，参量，决定因素
technologically	[teknə'lɔdʒikli]	adv. 技术上的

Part Ⅰ Windows——The most popular operating system

The first version of Microsoft Windows (Microsoft Windows 1.0) came out in November 1985. It had a graphical user interface, inspired by the user interface of the Apple computers of the time. Windows 1.0 was not successful with the public.②

1. Terms of Windows

表 7-1 Windows 操作系统的常用名词

Word	Chinese	Word	Chinese
desktop	桌面	Task bar	任务栏
icon	图标	menu	菜单
folder	文件夹	title bar	标题栏

续表

Word	Chinese	Word	Chinese
file	文件	directory	目录
copy	复制	paste	粘贴
cut	剪切	clipboard	剪贴板
button	按钮	default	默认
tool bar	工具栏	dialog box	对话框
option	选项	share	共享
boot	启动	user	用户
device	设备	server	服务器
host	主机		

2. Windows common error message

Windows 操作系统的常见错误提示如表 7-2 所示。

表 7-2 Windows 操作系统的常见错误提示

Hints and Tips	Chinese
CMOS battery failed	CMOS 电池失效
CMOS check sum error-Defaults loaded	CMOS 执行全部检查时发现错误，要载入系统预设值
Hard disk install failure	硬盘安装失败
Hard disk(s) diagnosis fail	执行硬盘诊断时发生错误
Hardware Monitor found an error，enter POWER MANAGEMENT SETUP for details，Press F1 to continue，DEL to enter SETUP	监视功能发现错误，进入 POWER MANAGEMENT SETUP 查看详细资料，或按 F1 键继续开机程序，按 Del 键进入 CMOS 设置
Press ESC to skip memory test	正在进行内存检查，可按 Esc 键跳过
Resuming from disk，Press TAB to show POST screen	从硬盘恢复开机，按 Tab 键显示开机自检画面
Abort, Retry, Ignore, fail?	退出，重试，忽略，取消？
Cannot find system files	未找到系统文件
Drive Not Ready Error	驱动器未准备好
Keyboard error or no keyboard present	键盘错误或者未接键盘

Part Ⅱ Mac OS——A operating system only good for graphics/ media work

Mac OS X is the most technologically advanced operating system Apple has ever released, but don't let that scare you. While there's a lot of powerful stuff going on under the hood, the Mac OS makes it easy for you to work, play, and get entertainment on your Mac.③

Mac OS command

Mac OS X 操作系统的常用命令如图 7-3 所示。

表 7-3 Mac OS X 操作系统的常用命令

Command	Function
cp	复制文件
rm	删除文件
mv	移动文件
chmod	更改文件权限
chown	更改文件属主
nano	文本编辑
sh	运行脚本命令

Part Ⅲ   Linux——A free operating system

Linux is fit for use on workstations as well as on middle-range and high-end servers. Today, a lot of the important players on the hardware and software market each have their team of Linux developers.④

1. Linux system command

Linux 操作系统的常用命令如表 7-4 所示。

表 7-4 Linux 操作系统的常用命令

Command	Function
insmod (install module)	载入模块
cd (change directory)	改变目录路径
cp (copy)	复制文件
find	查找文件
ps (process status)	进程状态，类似于 Windows 的任务管理器
rpm (redhat package manager)	红帽子打包管理器
su (switch user)	切换用户
more	分页显示

2. Linux common error message

Linux 操作系统的常见错误提示如表 7-5 所示。

表 7-5　Linux 操作系统的常见错误提示

Error Message	Chinese
command not found	没有发现命令
access denied	拒绝访问
general error	通常错误
invalid directory	不可用目录
syntax error	语法错误
required parameter missing	参数缺失
write protect error	写保护错误

Part Ⅳ　UNIX——A different operating system

UNIX is a multi-user（more than one user can use the machine at a time）, multi-tasking（more than one program can be run at a time ）operating system. Multiple users may have multiple tasks running simultaneously. This is very different than PC operating systems.[⑤]

1. UNIX system command

UNIX 操作系统的常用命令如表 7-6 所示。

表 7-6　UNIX 操作系统的常用命令

Command	Function
pwd　(print working directory)	显示出用户当前工作目录的路径名
mkdir　(make directory)	建立新目录
rmdir　(remove directory)	删除不存在文件的子目录名
ln　(link)	为文件或目录建立一个链
mv　(move)	改变文件或目录名，或把一些文件移到另一目录下
chown (change owner)	改变文件或目录的所有权
cmp (compare)	显示比较两文件不同处的信息

2. UNIX common error message

UNIX 操作系统的常见错误信息如表 7-7 所示。

表 7-7　UNIX 操作系统的常见错误信息

Command	Chinese
boot not found cannot open stage 1 boot failure: error loading hd (40) /boot.	表明系统根目录下的 Boot 文件丢失或找不到
unix not found	表明找不到 UNIX 操作系统
NO OS (Operating System)	此前系统能正常工作，表明系统硬盘的引导块被破坏

Exercises

1. Thinking.

在图 7-1 中，四张图片分别代表哪个操作系统？

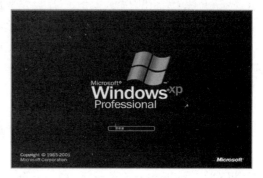

图 7-1　四种操作系统的界面

2. Translation.

（1）Press ESC to skip memory test

（2）Abort, Retry, Ignore, fail?

（3）Keyboard error or no keyboard present

（4）Resuming from disk，Press TAB to show POST screen

3. 简述 Windows、Mac OS、Linux、UNIX 四个操作系统的特点。

参　考　译　文

①　操作系统（Operating System，OS）是一个管理计算机硬件与软件的程序。它负责管理、调度、指挥计算机的软、硬件资源并使其协调工作。没有它，任何计算机都无法正常运行。它在资源使用者和资源之间充当中间人的角色。在计算机的发展过程中，出现过许多不同的操作系统，如 DOS，Mac OS，Windows，Linux，UNIX 等。

② Microsoft Windows 1.0 是 Microsoft 公司在 1985 年 11 月发布的第一代窗口式多任务系统。它有一个图形用户界面，受当时的苹果计算机的用户界面启发而得到。Microsoft Windows 1.0 并没能取得公众的认可。

③ Mac OS X 是苹果公司推出的技术最为先进的操作系统，但不要让它吓唬到。虽然有很多强大的程序需要在后台运行，但是 Mac OS 会使你的工作和游戏更为轻松，同时还能获取 Mac 为你提供的娱乐。

④ Linux 操作系统一般应用于工作站、中端和高端服务器。今天，许多硬件和软件市场上的龙头企业都拥有自己的 Linux 开发团队。

⑤ UNIX 是一个多用户（可供多个用户同时使用）、多任务（可同时运行多个程序）操作系统。多个用户可同时运行多个任务，这是它和其他计算机操作系统最大的区别。

CHAPTER | 08

Application Software
（应用软件）

As term suggests, application software such as Microsoft Office and WPS, provides certain special functions. They usually operate on operating systems (like Windows XP), and are developed by professional staffs to meet different needs. Most of the software we use are application software. For example: word processing software, webpage designing software, and animation software.

Lesson 1　Microsoft Office
（微软办公室配套软件）

❀━ Interrelated Knowledge

　　Microsoft Office 是微软公司开发的一套基于 Windows 操作系统的办公套装软件。其常用组件有 Word、Excel、Access、Powerpoint、FrontPage 等。目前其最新版本为 Office 2010。

　　Word 是文字处理软件。它被认为是 Office 的主要程序。它在文字处理软件市场上拥有统治份额。它私有的 DOC 格式被尊为一个行业的标准，虽然它的最新版本 Word 12.0/2007 也支持一个基于 XML 的格式。

　　Excel 是另一个重要组件，称它为电子表格软件，它主要用在数据管理、数据计算、数据分析、以图表说明数据等方面。Excel 中的函数更是为我们快速地对多个数据进行简单或复杂的计算提供了极大的方便。

　　Excel 中的函数其实是一些预定义的公式，它们使用一些叫做参数的特定数值按特定的顺序或结构进行计算。我们可以直接用它们对某个区域内的数值进行一系列运算，如分析和处理日期值和时间值、确定贷款的支付额、确定单元格中的数据类型、计算平均值、排序显示和运算文本数据等。函数还可以是多重的，也就是说一个函数可以是另一个函数的参数，这就是嵌套函数的含义。

❀━ Words and Expressions

function	['fʌŋkʃən]	n. 功能，函数　vi. 运行，起作用
sum	[sʌm]	n. 总数，金额　v. 总计，概括
average	['ævəridʒ]	n. 平均数，平均水平
number	['nʌmbə]	n. 号码，数字　vi. 总计，编号　vt. 编号
count	[kaunt]	v. 计算，视为，依赖　n. 计数，总数
insert	[in'sə:t]	v. 插入，嵌入　n. 插入物
hyperlink	['haipəlink]	n. 超链接

Part Ⅰ　An Introduction to Functions

Excel 中的部分函数列表如表 8-1 所示。

表 8-1　Excel 中的部分函数列表

Type	Function	Function Name
统计函数	对数据区域进行统计分析	Average，Count，CountA，CountIF，Max，MaxA，Min，MinA，Var
数学与三角函数	处理简单的计算	ABS，Acos，Asin，Atan，Cos，Exp，Int，Ln，Log，Mod，Odd，PI，Rand，Round，Sign，Sin，Sqrt，Sum，Tan
日期与时间函数	分析和处理日期值和时间值	Date，Day，Hour，Month，Now，Second，Time，Today，Weekday，Year

续表

Type	Function	Function Name
财务函数	进行一般的财务计算	DB, Ddb, Fv, Ipmt, Irr, Ispmt, Mirr, Npv, Pmt, Pv, Rate, Sln, Syd
逻辑函数	进行真假值判断,或复合检验	And, False, If, Not, Or, True
数据库函数	分析数据清单中的数值是否符合特定条件	Daverage, Dcount, Dget, Dmax, Dmin, Dprouduct, Dsum, Dvar, Dvarp
查询与引用函数	查找特定数值,或某一单元格的引用	Address, Areas, Choose, Column, Hlookup, Hyperlink, Index, Match, Offset, Row, Rows, Rtd, Transpose, Vlookup
信息函数	确定存储在单元格中的数据的类型	Cell, InFo, Isblank, Iserr, Iserror, Islogical, Isna, Isref, Istext, N, Type
文本函数	在公式中处理文字串	Asc, Char, Clean, Code, Exact, Find, Left, LeftB, Len, LenB, Lower, Mid, Midb, Proper, Replace, Right, RightB, Text, Trim, Upper, Value

函数的结构如图 8-1 所示,插入函数列表如图 8-2 所示。

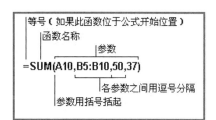

图 8-1 函数的结构

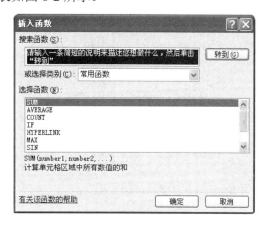

图 8-2 插入函数列表

Part Ⅱ Application of Function

1. Example for writing function

函数应用方法一如图 8-3 所示。

2. Example for inserting function

函数应用方法二如图 8-4 所示。
在"函数参数"框中选择和确认参数值的示意图如图 8-5 所示。

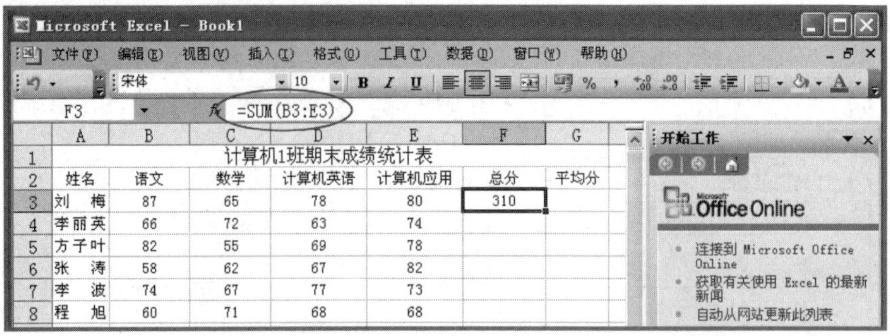

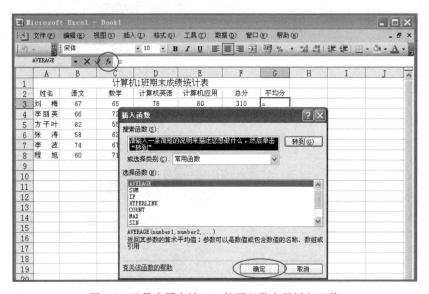

图 8-3　函数应用方法一：在编辑栏中直接输入函数及参数

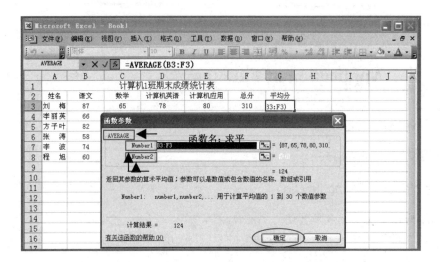

图 8-4　函数应用方法二：使用函数向导插入函数

图 8-5　在"函数参数"框中选择和确认参数值

Part Ⅲ   Commonly Used Functions in Excel

Excel 中的常用函数如表 8-2 所示。

表 8-2   Excel 中的常用函数

Name	Function	Format	Parameter Specification
Average	返回其参数的算术平均值	Average（number1,number2,……）	参数可以是数值或包含数值的名称、数组或引用
Sum	计算单元格区域中所有数值的和	Sum（number1,number2,…）	单元格中的逻辑值和文本将被忽略。但当作为参数输入时，逻辑值和文本有效
If	判断一个条件是否满足，如果满足返回一个值，如果不满足则返回另一个值	If（logical_test,value_if_true,value_if_false）	logical_test：任何一个可判断为 true 或 false 的数值或表达式 value_if_true：当 logical_test 为 true 时的返回值。如果忽略，则返回 true。If 函数最多可嵌套七层 value_if_false：当 logical_test 为 false 时的返回值。如果忽略，则返回 false
Hyperlink	创建一个快捷方式或链接，以便打开一个存储在硬盘、网络服务器或 Internet 上的文档	Hyperlink（link_location,friendly_name）	link_location：要打开的文件名称及完整路径。可以是本地硬盘、UNC 路径或 URL 路径 friendly_name：要显示在单元格中的数字或字符串。如果忽略此参数，单元格中显示 link_location 的文本
Count	计算包含数字的单元格以及参数列表中的数字的个数	Count（value1,value2,……）	可以包含或引用各种不同类型数据的参数，但只对数字型数据进行计数
Max	返回一组数值中的最大值，忽略逻辑值及文本	Max（number1,number2,……）	数值、空单元格、逻辑值或文本数值
Sin	返回给定角度的正弦值	Sin（number）	以弧度表示的角度
Sumif	对满足条件的单元格求和	Sumif（range,criteria,sum_range）	range：要进行计算的单元格区域 Criteria：以数字、表达式或文本形式定义的条件 sum_range：用于求和计算的实际单元格。如果省略，将使用区域中的单元格
Pmt	计算在固定利率下，贷款的等额分期偿还额	Pmt（rate,nper,pv,fv,type）	rate：各期利率 nper：总投资期或贷款期，即该项投资或贷款的付款期总数 pv：从该项投资（或贷款）开始计算时已经入账的款项，或一系列未来付款当前值的累积和 fv：未来值，或在最后一次付款后可以获得的现金余额。如果忽略，则认为此值为 0 type：逻辑值 0 或 1，用以指定付款时间在期初还是在期末。如果为 1，付款在期初；如果为 0 或忽略，付款在期末
Stdev	估算基于给定样本的标准偏差（忽略样本中的逻辑值及文本）	Stdev（number1,number2,……）	与总体抽样样本相应的数值，也可以是包含数值的引用

Lesson 2　Macromedia　Dreamweaver
（Macromedia 公司的 Dreamweaver）

❀ Interrelated Knowledge

　　Dreamweaver 是美国 Macromedia 公司开发的集网页制作和管理网站于一身的所见即所得网页编辑器，它是第一套针对专业网页设计师特别发展的视觉化网页开发工具。利用它可以轻而易举地制作出跨越平台限制和跨越浏览器限制的充满动感的网页。它支持最新的 Web 技术，包含 HTML 检查、HTML 格式控制、HTML 格式化选项、HomeSite/BBEdit 捆绑、可视化网页设计、图像编辑、全局查找替换、全 FTP 功能、处理 Flash 和 Shockwave 等富媒体格式和动态 HTML、基于团队的 Web 创作。Dreamweaver、Flash、Fireworks 三者被 Macromedia 公司称为 Dreamteam（梦之队）。

❀ Words and Expressions

tag	[tæg]	n. 标签，附属物
inspector	[in'spektə]	n. 检查员，巡视员
property	['prɔpəti]	n. 性质；财产
background	['bækgraund]	n. 背景，幕后
paragraph	['pærəgrɑ:f]	n. 段落，节　vt. 将……分段
marquee	[mɑ:'ki:]	n. 大天幕，华盖
embed	[im'bed]	vt. 使插入，使嵌入　vi. 嵌入

Part Ⅰ　Commonly Used HTML Tag

HTML 中的常用标签如表 8-3 所示。

表 8-3　HTML 中的常用标签

Tag name	Chinese	Tag name	Chinese
heading	标题	body	网页主体
link	链接	paragraph	段落
div	层	image	图像
list	列表	form	表单
input	表单的输入元素	frame	框架
span	范围	table	表格

Part Ⅱ   Dreamweaver tag inspector

1. Enter the tag inspector

进入 Dreamweaver 标签检查器界面的过程如图 8-6 所示。

图 8-6   进入 Dreamweaver 标签检查器的界面

2. Example for tag inspector

标签检查器的应用举例如图 8-7 所示。

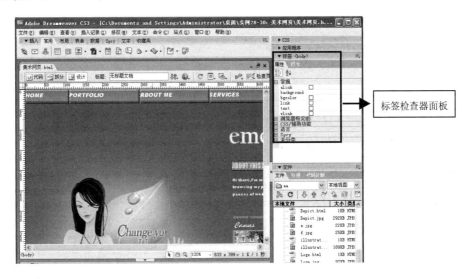

图 8-7   用标签检查器的常规选项中的"background"和"bgcolor"来对页面的背景和背景颜色进行设置

3. More tags

更多的 Dreamweaver 标签如图 8-8 所示。

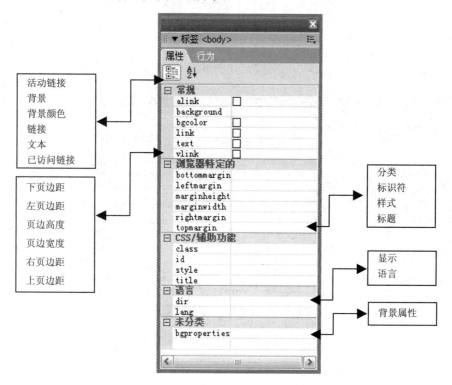

图 8-8　更多的 Dreamweaver 标签

Part Ⅲ　Dreamweaver HTML Code

1. Properties of the HTML tag

HTML 常用标签的属性如表 8-4 所示。

表 8-4　HTML 常用标签的属性

Paragraph（段落标签常用属性）		Table（表格标签常用属性）	
font-family	字体	cellpadding	单元格边距
font-weight	粗细	cellspacing	单元格间距
font-style	样式	border	边框
font-size	大小	width	宽度
align	文本对齐	height	高度
color	颜色	background	背景
Marquee（滚动标签常用属性）		Embed（嵌入标签常用属性）	
scrollamount	滚动速度	autostart	自动播放
direction	滚动方向	loop	循环
scrolldelay	延时时间	volume	音量

2. Example for table tag

table 标签的应用举例如图 8-9 所示。

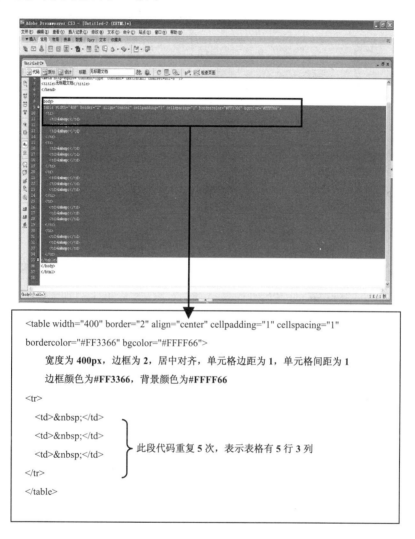

图 8-9　运用 Table 标签在 Dreamweaver 中插入一个 5 行 3 列的表格

3. Example for embed tag

Embed 标签的应用举例如图 8-10 所示。

4. Example for marquee tag

Marquee 标签的应用举例如图 8-11 所示。

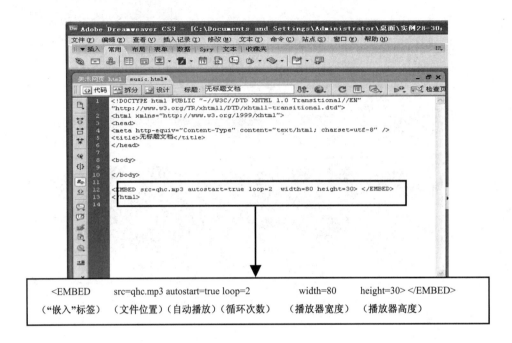

图 8-10　运用 Embed 标签在 Dreamweaver 中插入名为 qhc.mp3 的背景音乐

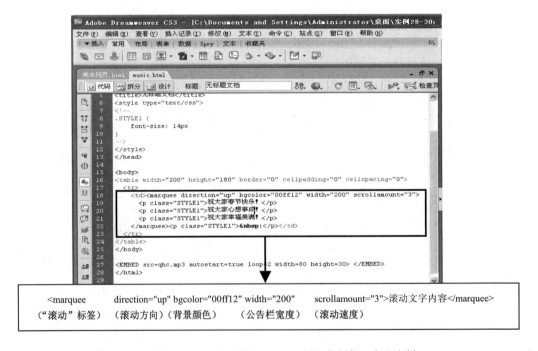

图 8-11　运用 Marquee 标签在 Dreamweaver 中制作一个公告栏

Lesson 3　Macromedia Flash
（Macromedia 公司的 Flash 软件）

❀ Interrelated Knowledge

　　Flash 是美国 Macromedia 公司所设计的一种二维动画软件。它通常包括 Macromedia Flash（用于设计和编辑 Flash 文档），以及 Macromedia Flash Player（用于播放 Flash 文档）。现在 Flash 已经被 Adobe 公司购买，Flash 影片的后缀名为 .swf，该类型文件必须用 Flash 播放器才能打开，但由于其占用硬盘空间少，所以现在被广泛应用于游戏和网页动画中。Flash 软件可以实现多种动画特效（动画都是由一帧帧的静态图片在短时间内连续播放而造成的视觉效果），是表现动态过程、阐明抽象原理的一种重要媒体。

❀ Words and Expressions

familiar	[fə'miljə]	adj. 熟悉的，熟知的
action	['ækʃən]	n. 动作，行动
script	[skript]	n. 脚本，原稿，手稿，手迹
root	[ru:t]	n. 根，根源，根本
volume	['vɔljum]	n. 卷，册，容量，音量

Part Ⅰ　Familiar Word in Flash

Flash 中常用的英文单词如表 8-5 所示。

表 8-5　Flash 中常用的英文单词

Word	Chinese	Word	Chinese
frame	帧	movieclip	影片剪辑
scene	场景	stage	舞台
button	按钮	variable	变量
event	事件	library	库
symbol	符号	timer	计时器
load	加载	press	按下
release	松开	duplicate	复制
import	导入	sound	声音
rotation	旋转	scale	范围
visible	可见性	random	随机
drag	拖曳	roll	滚动

Part Ⅱ ActionScript

1. Enter Flash ActionScript Panel

进入 Flash 动作脚本界面的步骤如图 8-12 所示。

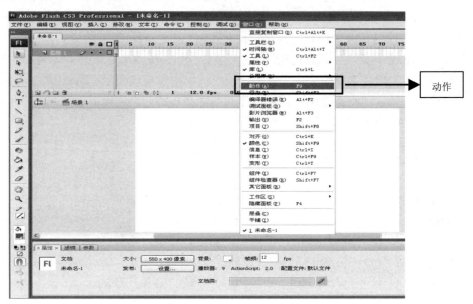

图 8-12　进入 Flash 动作脚本的界面

2. Example for ActionScript

Flash 动作脚本的应用举例如图 8-13 和图 8-14 所示。

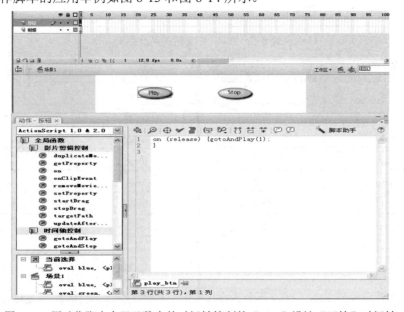

图 8-13　用动作脚本全局函数中的时间轴控制的"play"设计"开始"时间轴

CHAPTER 08　Application Software（应用软件）

图 8-14　用动作脚本全局函数中的时间轴控制的"stop"设计"停止"时间轴

3. Familiar ActionScript Type

时间轴函数如图 8-15 所示。

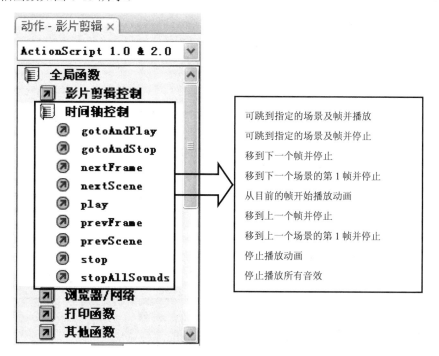

图 8-15　时间轴函数

影片剪辑控制函数如图 8-16 所示。

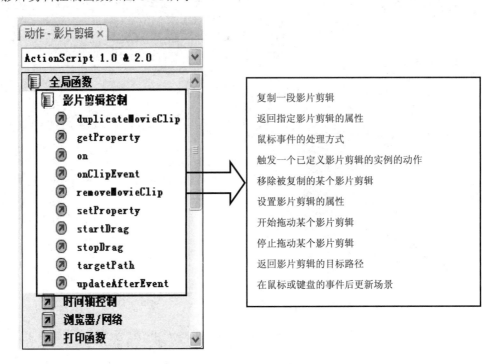

图 8-16　影片剪辑控制函数

浏览器/网络函数如图 8-17 所示。

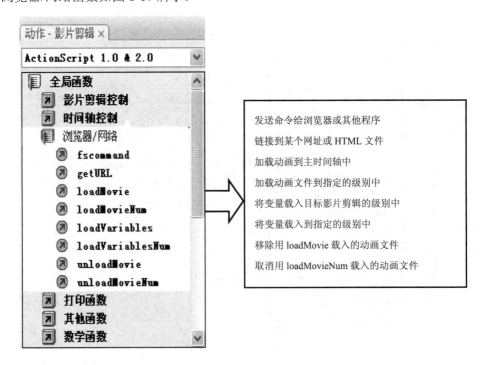

图 8-17　浏览器/网络函数

Part Ⅲ Familiar ActionScript code in Flash

1. 声音加载

mysound=new Sound();	//创建声音实例
mysound.attachSound("music");	//将音乐 music 链接到 sound 实例
mysound.start(0,2);	//播放音乐 2 遍
mysound.setVolume(70);	//音乐的音量为 70%

2. 鼠标跟随

n= Number(n)+24;	//n 的数值每次递增 24
if (Number(n)<360) {	//当 n 小于 360 时
duplicateMovieClip("movie", "movie"add n, n);	//复制动画片段，并命名为 movie 以深度 n 复制
setProperty ("_root.movie"add n, _rotation, getProperty("movie",_rotation)–n);	//取得 movie 的旋转角度值，将其减去 n，并将所得值设为目标"_root.movie" add n 的旋转角度值。
gotoAndPlay(1);	//跳到第 1 帧
}	

Exercises

1. Match.

Average	A．求和函数
Sum	B．计数函数
Count	C．求最大值函数
Max	D．求最小值函数
Min	E．求平均值函数

2. Translation.

marquee
duplicate
frame
stage
tag

3．Thinking.

请写出以下代码将要建立怎样的表格？

```
<table width="400" border="2" align="center" cellpadding="1" cellspacing="2" bordercolor="#3333FF"    bgcolor="#FF99FF ">
   <tr>
     <td> </td>
     <td> </td>
     <td> </td>–
   </tr>
   <tr>
     <td> </td>
     <td> </td>
     <td> </td>
   </tr>
   <tr>
     <td> </td>
     <td> </td>
     <td> </td>
   </tr>
</table>
```

参 考 译 文

顾名思义，应用软件即是提供某种特定功能的软件，如 Office、WPS 等，它们一般都运行在操作系统（如 Windows XP）之上，由专业人员根据各种需要开发。我们平时见到和使用的大部分软件均为应用软件，如文字处理软件、网页制作、动画制作软件等。

CHAPTER 09

Tool Software
（常用工具软件）

What is known as "Tool Software" is what helps computer users with auxiliary efficiency in various fields, and in the meantime assisting the application software. Common ones include cartographic software, antivirus software, and imprinting software.

Lesson 1　Compression and Decompression Software
（解压缩软件）

Interrelated Knowledge

文件压缩是指将原文件格式中的冗余空字符进行合并，并记录下来；解压缩就是指将压缩文件还原成原文件。各种不同的压缩软件的压缩比是不一样的，而对同一个软件也可以有不同的压缩比。WinRAR 是目前流行的压缩工具，其界面友好，使用方便，在压缩率和速度方面都有很好的表现。其压缩率比高，3.x 采用了更先进的压缩算法，是现在压缩率较大、压缩速度较快的格式之一。3.3 增加了扫描压缩文件内病毒和解压缩 zip 文件的功能，升级了分卷压缩的功能等。

Words and Expressions

profile	['prəufail]	n. 概要
split	[split]	v. 分离，分开
decompress	[di:kəm'pres]	vt. 缓缓排除压力，减压
destination	[desti'neiʃən]	n. 目的地，终点
extraction	[iks'trækʃən]	n. 抽出，取出，抽出物
folder	['fəuldə]	n. 文件夹，折叠者
prompt	[prɔmpt]	vi. 提示　n. 提示，提示的内容
authenticity	[ɔ:θen'tisiti]	n. 确实性，真实性
miscellaneous	[misi'leinjəs]	n. 杂货，杂项
subfolder	['sʌbfəuldə]	n. 子文件夹

Part Ⅰ　Interface of WinRAR

WinRAR 的用户界面如图 9-1 所示。

Part Ⅱ　How to use WinRAR

1. Compressed files

压缩文件的工作界面如图 9-2 所示。

> Archiving options（压缩选项）
> Delete files after archiving：压缩后删除源文件。
> Create SFX archive：创建自解压格式压缩文件。
> Create solid archive：创建固实压缩文件。

Put authenticity verification：添加用户身份校验信息。
Put recovery record：添加恢复记录。
Test archived files：测试压缩文件。
Lock archive：锁定压缩文件。

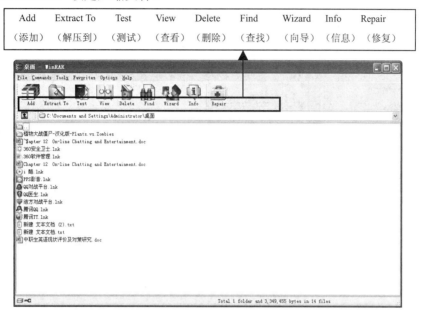

图 9-1　WinRAR 的用户界面

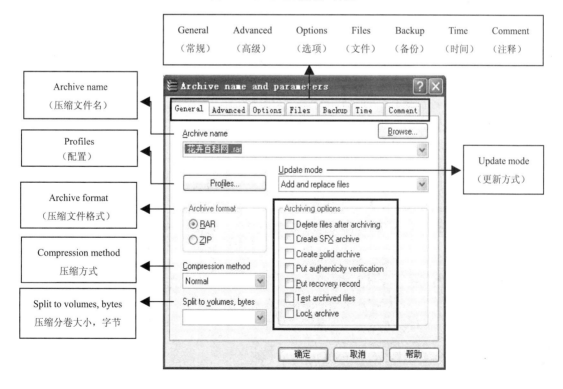

图 9-2　压缩文件的工作界面

2. Decompressing files

解压文件的工作界面如图 9-3 所示。

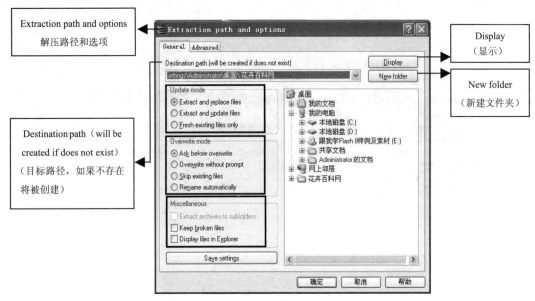

图 9-3　解压文件的工作界面

- Update mode（更新方式）
  Extract and replace files：解压并替换文件。
  Extract and update files：解压并更新文件。
  Fresh existing files only：仅更新已经存在的文件。
- Overwrite mode（覆盖方式）
  Ask before overwrite：覆盖前询问。
  Overwrite without prompt：没有提示直接覆盖。
  Skip existing files：跳过已经存在的文件。
  Rename automatically：自动重命名。
- Miscellaneous（其他）
  Extract archives to subfolders：解压缩文件到子文件夹。
  Keep broken files：保留损坏的文件。
  Display files in Explorer：在资源管理器中显示文件。

Lesson 2　Picture Browsing Software（图片浏览软件）

Interrelated Knowledge

ACDSee 是一款重量级的图像浏览软件，它不仅能用于图片浏览，同时还有图像获取、管理、优化甚至和他人分享的功能，因此我们可以从数码相机或者扫描仪获取图片并利用

ACDSee 对其进行便捷的查找、组织和预览。需要提及的是，ACDSee 能快速显示所需的图片，而且能够以内置的播放器播放 MPEG 之类的影像文件。另外，ACDSee 还具有图片编辑功能，可以对图片进行去红眼、剪切、锐化、浮雕特效、曝光调整、旋转、镜像等特效处理，让我们的图像更具真实感。

Words and Expressions

filter	['filtə]	n. 筛选　　v. 过滤，渗透
sort	[sɔ:t]	n. 种类　　v. 分类，整理，使明确
album	['ælbəm]	n. 相册，集邮册
various	['vɛəriəs]	adj. 各种各样的
pending	['pendiŋ]	adj. 待定的，即将发生或来临的

ACDSee 的用户界面如图 9-4 所示。

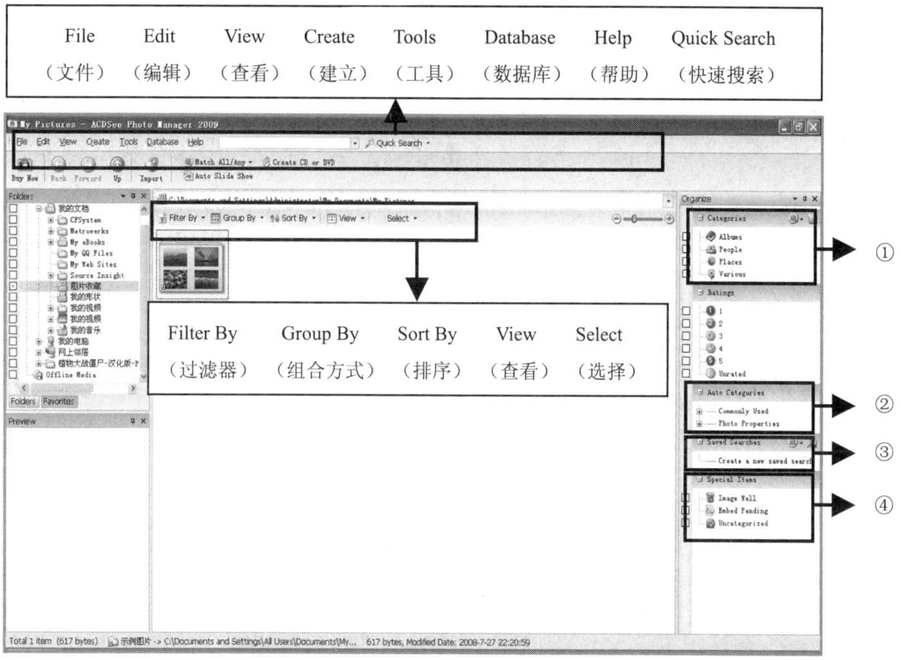

图 9-4　ACDSee 的用户界面

➢ ① Categories（类别）
Album：　相册。
People：　人物。
Place：　地点。
Various：　花朵。

➢ ② Auto Categories（自动分类）
Commonly Used：　常用。
Photo Properties：　相片参数。

➢ ③ Save Searches（保存搜索）
Create a new save search
创建一个新的保存搜索。

➢ ④ Special Items（特殊项目）
Image Well：　图像库。
Embed Pending：　嵌入等待。
Uncategorized：　未分类。

Lesson 3　AntiVirus Software
（杀毒软件）

❀ Interrelated Knowledge

　　计算机病毒是一个程序，一段可执行码。就像生物病毒一样，计算机病毒有独特的复制能力。计算机病毒可以很快地蔓延，又常常难以根除。它们能把自身附着在各种类型的文件上。当文件被复制或从一个用户传送到另一个用户时，它们就随同文件一起蔓延开来。

　　除复制能力外，某些计算机病毒还有其他一些共同特性：一个被污染的程序能够传送病毒载体。当你看到病毒载体似乎仅仅表现在文字和图像上时，它们可能已毁坏了文件，格式化了你的硬盘驱动或引发了其他类型的灾害。若是病毒并不寄生于一个污染程序，它仍然能通过占据存储空间给你带来麻烦，并降低你的计算机的全部性能。

　　世界著名的杀毒软件有麦咖啡（McAfee）、诺顿（Norton）和卡巴斯基（Kaspersky）等，国产著名杀毒软件有江民、瑞星和金山。杀毒软件应具备三项基本功能：查毒杀毒功能；防毒（即实时防护）功能；更新升级功能。杀毒软件病毒库要定期更新，因为计算机病毒层出不穷，如一些新兴的具有蠕虫特性的病毒具有相当快的传播速度和繁殖能力，陈旧的病毒库无法对付新病毒。只有保持病毒库的经常更新，才能保证杀毒软件充分发挥作用。

❀ Words and Expressions

anti	['ænti]	adj. 反对的
virus	['vaiərəs]	n. 病毒
anti-virus software		防病毒软件
protection	[prə'tekʃən]	n. 保护，防护
worm	[wə:m]	n. 虫，蠕虫
macro	['mækrəu]	n. 宏
Trojan	['trəudʒən]	n. 特洛伊
backdoor	['bækdɔ:]	adj. 秘密的
joke	[dʒəuk]	n. 笑话，玩笑
hacker	['hækə]	n. 电脑黑客
binder	['baində]	n. 捆绑机

Part Ⅰ　Virus

　　常见病毒名称前缀表如表 9-1 所示。

表 9-1 常见病毒名称前缀表

Prefix	Meaning	Explanation
Worm	蠕虫病毒	蠕虫病毒的公有特性是通过网络或者系统漏洞进行传播，很大部分的蠕虫病毒都有向外发送带毒邮件，阻塞网络的特性，如冲击波（阻塞网络）、小邮差（发带毒邮件）等
Trojan	特洛伊木马	特洛伊木马会通常假扮成有用的程序诱骗用户主动激活，或利用系统漏洞侵入用户计算机。木马进入用户计算机后会首先隐藏在系统目录下，然后修改注册表，完成黑客制定的操作，如 Trojan.Win32.PGPCoder.a（文件加密机）、 Trojan.Reper.i（推销员变种 I）
Hack	黑客病毒	黑客病毒有一个可视的界面，能对用户的计算机进行远程控制。木马、黑客病毒往往是成对出现的，即木马病毒负责侵入用户的计算机，而黑客病毒则会通过该木马病毒来进行控制。现在这两种类型都越来越趋向于整合了
Backdoor	后门病毒	后门病毒会通过网络或者系统漏洞进入用户的计算机并隐藏在系统目录下，被开后门的计算机可以被黑客远程控制，如 Backdoor.Huigezi.ik （灰鸽子变种 IK）、 Backdoor.Heidong （黑洞）
Macro	宏病毒	宏病毒是使用宏语言编写的，可以在一些数据处理系统中运行（主要是微软的办公软件系统，字处理、电子数据表和其他 Office 程序中）。 有时会根据具体情况详细标明是针对何种处理系统的病毒。Macro.Word 代表感染 Word 文档的病毒， Macro.Excel 代表感染 Excel 文档的病毒，如 Macro.Word.Apr30 （四月三十宏病毒）
Joke	恶作剧程序	它不会对用户的计算机、文件造成破坏，但可能会给用户带来恐慌和不必要的麻烦，如女鬼（Joke.Girlghost）病毒
Win32 PE Win95 W32 W95	系统病毒	这些病毒的一般公有特性是可以感染 Windows 操作系统的 *.exe 和 *.dll 文件，并通过这些文件进行传播，如大家熟悉的 CIH 病毒
Script VBS JS	脚本病毒	脚本病毒的公有特性是使用脚本语言编写，通过网页进行传播的，如红色代码（Script.Redlof）。脚本病毒还会有如下前缀：VBS、JS（表明是用何种脚本编写的），如欢乐时光（VBS.Happytime）、十四日（Js.Fortnight.c.s）等
Binder	捆绑机病毒	这类病毒的公有特性是病毒与一些应用程序如 QQ、IE 捆绑起来，当用户运行了捆绑病毒的应用程序时，会给用户造成危害，如捆绑 QQ（Binder.QQPass.QQBin）、系统杀手（Binder.killsys）等

Part Ⅱ Scan Viruses

1. Interface of Protection Center

诺顿软件主界面—防护中心如图 9-5 所示。

2. Interface of Norton AntiVirus

诺顿软件主界面——诺顿杀毒如图 9-6 所示。

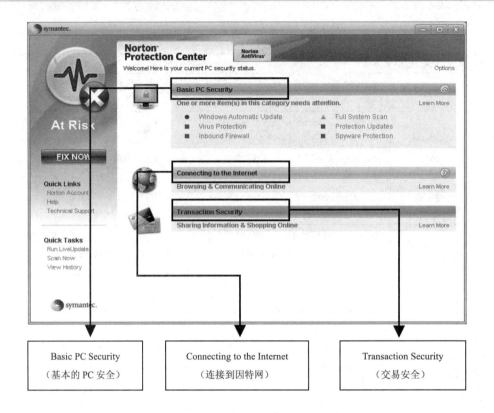

图 9-5　诺顿软件主界面—防护中心

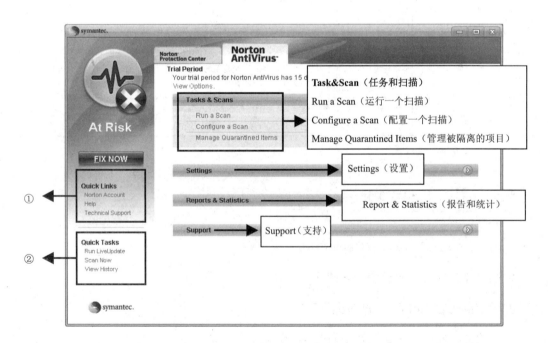

图 9-6　诺顿软件主界面—诺顿杀毒

➢ ① Quick Links（快速连接）
Norton：Account：诺顿账号。
Help：帮助。
Technical Support：技术支持。
➢ ② Quick Tasks（快速启动任务）
Run LiveUpdates：运行在线升级。
Scan Now：现在扫描。
View History：查看历史记录。

Part Ⅲ AntiVirus LiveUpdate

诺顿软件的在线升级如图 9-7～图 9-9 所示。

图 9-7 诺顿软件的在线升级 1

➢ Welcome to LiveUpdate
欢迎来到在线升级。

➢ The following Symantec products and components are installed on your computer:
您的计算机上安装了以下的 Symantec 产品和组件。

➢ LiveUpdate will use the Internet to search for updates your installed Symantec products and components.
在线升级将通过互联网搜索您安装的 Symantec 产品和组件的更新。

➢ Click Next to see available updates.
单击"下一步"按钮查找可获取的更新项。

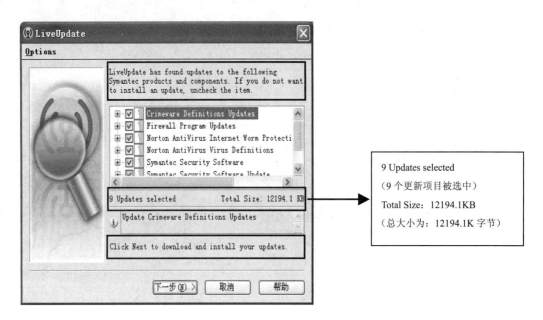

图 9-8　诺顿软件的在线升级 2

- ➢ LiveUpdate has found updates to the following Symantec products and components. If your do not want to install and update, uncheck the item.
  在线升级已经找到下列可更新的 Symantec 产品和组件。如果您不想安装某个更新，请取消对该项的选择。
- ➢ Click Next to download and install your updates.
  单击"下一步"按钮下载并安装您的更新选项。

图 9-9　诺顿软件的在线升级 3

> Thank you for using LiveUpdate. The following Symantec products and components are now up-to-date.
> 感谢您使用在线升级。下列 Symantec 产品和组件现在开始更新。
> LiveUpdate successfully download and installed this update.
> 在线升级已成功下载并安装了这次更新。

Lesson 4　Firewall——Agnitum Outpost Firewall（防火墙）

Interrelated Knowledge

病毒、木马、黑客都能对我们的计算机造成巨大的威胁。而随之衍生的杀毒软件、木马查杀、防火墙也渐渐成为我们在网络生活中的必备软件。杀毒软件和木马查杀软件主要用于处理计算机内部的问题，防火墙主要用于处理来自网络外的威胁，这样就更具主动防御功能。

Agnitum Outpost Firewall 口碑很好，用户对其评价很高，号称世界排名第二，它的功能是同类 PC 软件中比较强大的，包括了广告和图片过滤、内容过滤、DNS 缓存等功能。它能够预防来自 Cookies、广告、电子邮件病毒、后门、窃密软件、解密高手、广告软件和其他 Internet 危险的威胁，但是可能会占用系统资源多一些。该软件不需配置即可使用，这对许多新手来说很友好。

Words and Expressions

firewall	['faiəwɔ:l]	n. 防火墙
outpost	['autpəust]	n. 前哨，前哨基地，警戒部队
security	[si'kju:riti]	n. 安全；防护措施
block	[blɔk]	n. 阻塞（物），障碍（物）　　v. 阻塞

Outpost 软件主界面—防火墙如图 9-10 所示。

> Firewall is a key part of your system's protection. It monitors all the network traffic the system sends and receives, and it detects and prevents any hacker attacks from outside the network leaving intruders no possibility of compromising your computer.
> 防火墙是系统保护重要的一部分。它监测系统发送和接收的网络流量，探测和防止任何黑客从外部网络发动的攻击，使您的计算机免受损害。

> ① My Security（我的安全）
> Firewall：防火墙。
> Host Protection：主机保护。

> ② Event viewer（事件查看）
> Product Internal Events：产品内部事件。
> Firewall：防火墙。
> Anti-Leak Control：防漏控制。

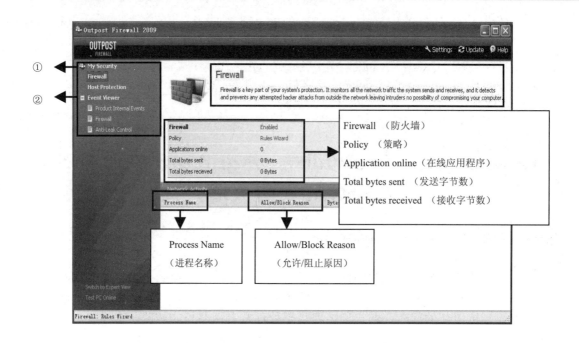

图 9-10 Outpost 软件主界面—防火墙

Lesson 5 Screen Capture Software（屏幕捕捉软件）

❀ Interrelated Knowledge

TechSmith SnagIt 是一个非常著名的优秀屏幕、文本和视频捕获、编辑与转换软件。它可以捕获 Windows 屏幕、DOS 屏幕；RM 电影、游戏画面；菜单、窗口、客户区窗口、最后一个激活的窗口或用鼠标定义的区域；可以选择是否包括光标，添加水印。另外，它还具有自动缩放、颜色减少、单色转换、抖动，以及转换为灰度级等功能。其图像可保存为 BMP、PCX、TIF、GIF 或 JPEG 格式，也可以存为视频动画。使用 JPEG 可以指定所需的压缩级别（从 1%到 99%）。

此外，SnagIt 在保存屏幕捕获的图像之前，还可以用其自带的编辑器进行编辑；也可选择自动将其送至 SnagIt 虚拟打印机或 Windows 剪贴板中，或直接用 E-mail 发送。

❀ Words and Expressions

launch	[lɔ:ntʃ]	v. 发射，发动，发起
capture	['kæptʃə]	vt. 捕获，（用照片等）留存
convert	['kɔnvə:t]	v. 变换，（使）转变
accessory	[æk'sesəri]	adj. 附属的（副的，辅助的）　n. 附件
scroll	[skrəul]	n. 卷轴，目录
delay	[di'lei]	v.推迟，延误　n.推迟，延期

SnagIt 软件主界面如图 9-11 所示。

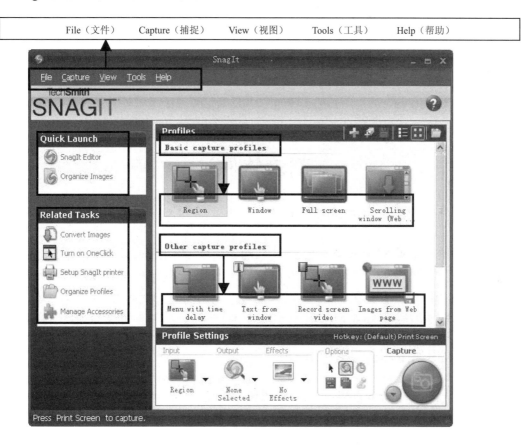

图 9-11　SnagIt 软件的主界面

➢ Quick Launch（快速启动）
　　SnagIt Editor：SnagIt 编辑器。
　　Organize Images：管理图像。
➢ Related Tasks（相关）
　　Convert Images：转换图像。
　　Turn on OneClick：打开一键操控。
　　Setup SnagIt printer：安装 SnagIt 打印机。
　　Organize Profile：管理方案。
　　Manage Accessories：管理插件。
➢ Basic capture profiles（基本捕捉方案）
　　Region：区域。
　　Window：窗口。
　　Full screen：全屏幕。
　　Scrolling window(Web)：活动窗口（网页）。

> Other capture profiles（其他捕捉方案）
> Menu with time delay：延时菜单。
> Text from window：窗口中的文字。
> Record screen video：录制屏幕视频。
> Image form Web page：网页中的图像。

Lesson 6　Burning Software
（刻录软件）

❀ Interrelated Knowledge

NERO-Burning Rom 是德国 ahead 公司出品的一款非常出色的刻录软件，容易使用，功能强大而齐全。它支持中文长文件名刻录，使用 Nero 可让您以轻松快速的方式制作您专属的 CD 和 DVD。不论您所要刻录的是资料 CD，音乐 CD，Video CD，Super Video CD，DDCD 或是 DVD，所有的程序都是一样的。

❀ Words and Expressions

burn	[bə:n]	v. 燃烧，烧着　　n. 燃烧
entertainment	['entə'teinmənt]	n. 娱乐
compile	[kəm'pail]	vt. 编译，编纂
compilation	[kɔmpi'leiʃən]	n. 编译，编辑
track	[træk]	n. 跑道，轨道，踪迹；　v. 跟踪，追
format	['fɔ:mæt]	n. 格式
favorite	['feivərit]	adj. 喜爱的　　n. 最喜爱的人或物，收藏夹

Part Ⅰ　Interface of Nero

Nero 软件的开始界面如图 9-12 所示。

> Complete your tasks quickly and easily.
> 快速轻松地完成任务。
> Manage your projects step-by-step with the application launcher that does it all. Special direct functions make it even easier!
> 使用应用程序启动程序即可对您的项目进行分步管理。特殊指示功能使管理变得更简单！

CHAPTER 09　Tool Software（常用工具软件）

图 9-12　Nero 软件的开始界面

Part Ⅱ　How to use the Nero

1. Data Burning

Nero 的数据刻录界面如图 9-13 所示。

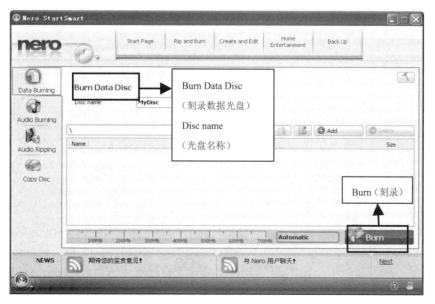

图 9-13　Nero 的数据刻录界面

2. Audio Burning

Nero 的音频刻录界面如图 9-14 所示。

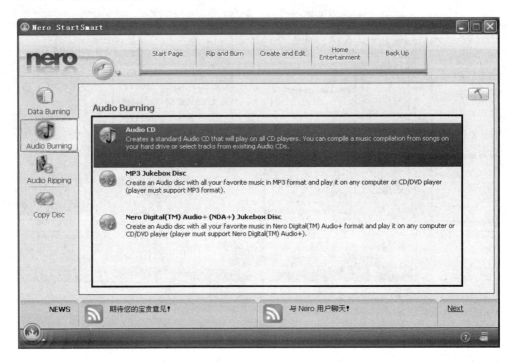

图 9-14 Nero 的音频刻录界面

➢ Audio CD（音频 CD）

Create a standard Audio CD that will play on all CD players. You can compile a music compilation from songs on your hard driver or select tracks from existing Audio CDs.

创建可在所有 CD 播放机播放的标准音频 CD。您可以根据硬盘上的歌曲对音乐内容进行编译，也可以选择现有音频 CD 中的轨道。

➢ MP3 Jukebox Disc（MP3 Jukebox 光盘）

Create an Audio disc with all your favorite music in MP3 format and play or CD/DVD player (MP3 format).

使用您最喜欢的 MP3 格式音乐创建音频光盘，并在任何计算机或 CD/DVD 播放机上播放（播放机必须支持 MP3 格式）。

➢ Nero Digital(TM) Audio+ (NDA+) Jukebox Disc（Nero Digital(TM) + (NDA+) Jukebox 光盘）

Create an Audio disc with all your favorite music in Nero Digital(TM) Audio+ format and it on any computer or CD/DVD player (player must support Nero Digital(TM) Audio+).

使用您最喜欢的 Nero Digital(TM) Audio+格式音乐创建音频光盘，并在任何计算机或 CD/DVD 播放机上播放（播放机必须支持 Nero Digital(TM) Audio+格式）。

Exercises

1. Blank.

backdoor	account	associate	folder	hacker
license	joke	verification	overwrite	macro
Trojan	authenticity	integration	worm	update

以上英文单词中代表病毒前缀的单词有：
_____、_____、_____、_____、_____

2. 请翻译（汉译英）
（1）解压缩软件。
（2）图片浏览软件。
（3）杀毒软件。
（4）刻录软件。
（5）屏幕捕捉软件。

参 考 译 文

所谓"工具软件"即帮助计算机用户在不同领域达到辅助效果，同时对应用程序起到辅助作用的软件。常用的工具软件包括制图软件、杀毒软件、刻录工具软件等。

第四篇

网络篇

CHAPTER 10

Computer Network
（计算机网络）

A computer network is a group of computers connected together to exchange or share information. Beginners like us can learn something about types of computer network and network hardware. There are many kinds of computer networks. In terms of network topology, there are bus, star, ring and tree networks. In terms of covered distance, there are three types of networks: LAN, MAN and WAN. In addition, there are many kinds of common hardware we should know about. Let's go and learn right now!

Lesson 1　　Basic Terminology
（基本术语）

🌸 Interrelated Knowledge

　　计算机网络，就是把分布在不同地理区域的计算机与专门的外部设备用通信线路互联成一个规模大、功能强的网络系统，从而使众多的计算机可以方便地互相传递信息，共享硬件、软件、数据信息等资源。如图 10-1 所示是办公室的计算机网络模型。

图 10-1　办公室的计算机网络模型

　　计算机网络按分界距离可分为三类：局域网（LAN）、城域网（MAN）、广域网（WAN）；按网络拓扑结构可分为四类：总线形网络、星形网络、环形网络、树形网络，如图 10-2 所示。

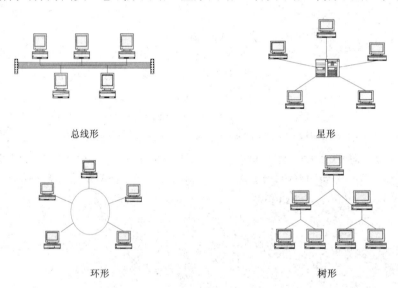

图 10-2　四种网络拓扑结构类型

🌸 Words and Expressions

network	['netwə:k]	n. 网络；vt. 联络，交流
connect	[kə'nekt]	v. 连接，接通
hardware	['hɑ:dwɛə]	n. （计算机）硬件
distance	['distəns]	n. 距离
bus	[bʌs]	n. [计]总线

ring	[riŋ]	n. 环，戒指
apart from this		除此之外
terminology	['tə:mi'nɔlədʒi]	n. 术语，术语学

Part Ⅰ　Basic Terminology of Network

计算机网络基本术语如表 10-1 所示。

表 10-1　计算机网络基本术语

Network Term	Chinese Meaning
PC （Personal Computer）	个人电脑，又称微型计算机或微机
LAN （Local Area Network）	局域网
MAN （Metropolitan Area Network）	城域网
WAN （Wide Area Network）	广域网
Ethernet	以太网
Workstations	工作站
Host	主机
Subnet	子网
Client / Server	客户端/服务器
Switch	交换机
MODEM	调制解调器
Router	路由器
Hub	集线器
NIC （Network Interface Card）	网卡
Gateway	网关
Bandwidth	带宽
Channel	信道
DNS （Domain Name System）	域名系统
IP （Internet Protocol）	网际协议
FTP （File Transfer Protocol）	文件传输协议
TCP （Transmission Control Protocol）	传输控制协议

Part Ⅱ　Common Terminology of Internet

互联网常见术语如表 10-2 所示。

表 10-2　互联网常见术语

Internet Term	Chinese Meaning
Internet（International Net）	因特网，国际互联网
WWW（World Wide Web）	万维网
HTTP（Hypertext Transfer Protocol）	超文本传输协议
BBS（Bulletin Board System）	论坛，电子公告板系统
E-mail（Electronic Mail）	电子邮件
banner ad.	横幅广告
LOGO	图标广告
hacker	黑客
blog	博客
user	用户
Netizen（net citizen）	网民
cyber friend	网友
download	下载
firewall	防火墙
broadband	宽带
Bug	漏洞，计算机内部发生的小故障
IE（Internet Explorer）	微软开发的浏览器
IT（Information Technology）	信息产业
SOHO（Small Office Home Office）	在家办公

Lesson 2　Common Kinds of Network Hardware（常用网络硬件）

Interrelated Knowledge

组成一般计算机网络的硬件有哪些呢？一是网络服务器；二是网络工作站；三是网络适配器，又称网络接口卡或网卡；四是连接线，学名为"传输介质"或"传输媒体"，主要是电缆或双绞线，还有不常用的光纤。如果要扩展局域网的规模，就需要增加通信连接设备，如调制解调器、集线器、网桥和路由器等。把这些硬件连接起来，再安装上专门用来支持网络运行的软件（包括系统软件和应用软件），那么一个能够满足工作或生活需求的计算机网络也就建成了（如图 10-3 所示）。

CHAPTER 10　Computer Network（计算机网络）

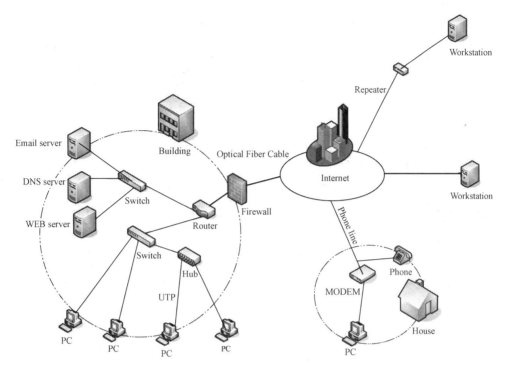

图 10-3　计算机网络模型

❀ Words and Expressions

server	['sə:və]	n. 服务器
modem	['məudem]	n. 调制解调器
repeater	[ri'pi:tə]	n.【计】中继器，转发器
optical	['ɔptikəl]	adj. 光学的，视觉的
fiber	['faibə]	n. 纤维（物质），力量
cable	['keibl]	n. 电缆

常用的网络硬件如表 10-3 所示。

表 10-3　常用的网络硬件

| Server（服务器） | Firewall（防火墙） | Workstation（工作站） | Switch（交换机） |

续表

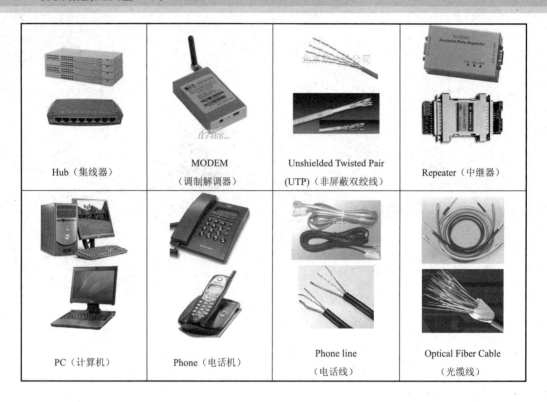

Exercises

1. Blank.

（1）在图 10-4 所示的网络示意图中，请指出标识的各种网络硬件的中文名称。

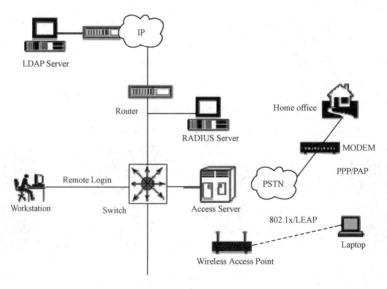

图 10-4　网络示意图 1

（2）图 10-5 是一个学校的网络示意图，请指出相应的网络硬件的英文名称。

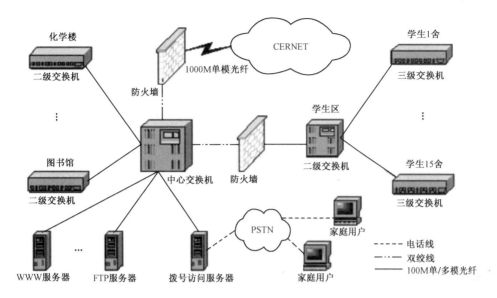

图 10-5　网络示意图 2

网络硬件名称（中文）——（英文）	网络硬件名称（中文）——（英文）

2. Understand the meaning of the following abbreviations and its full name in English.

PSTN
DNS
TCP
IP

参 考 译 文

简单地说，计算机网络就是将多台计算机连接起来。像我们这样的初学者可以先了解一些有关网络分类和网络硬件方面的知识。计算机网络可以分为很多种类型，如总线形网络、星形网络、环形网络和树形网络。当然计算机网络还可以根据联网的计算机之间的距离和网络覆盖面的不同分为局域网、城域网和广域网。除此之外，我们还要熟知一些常用网络硬件。让我们现在就开始学习吧！

CHAPTER 11

About Internet
(关于 Internet)

With the development of computer technology, the Earth Village is no longer a dream. We can retrieve any information form the Internet—a huge global computer networking system that contains a countless amount of shared information sources. It provides various kinds of online services to the world, including email, instant messaging, online shopping, videos, journals or even concerts. Internet is the inevitable outcome of computer network development.

Lesson 1　Primary Internet Services
（Internet 的主要服务）

❋ Interrelated Knowledge

Internet 提供了丰富的信息资源和应用服务。通过 WWW，人们可以非常方便地浏览、查询、下载、复制和使用这些信息；远程登录（Telnet 服务）和 FTP 服务能够让这些信息实现资源共享；而 E-mail、BBS、博客及网上聊天等服务能及时地传送文字、声音、图像等信息，实现信息交流。

❋ Words and Expressions

WWW（World Wide Web）		万维网
Web	[web]	n. 网页
Browser	['brauzə]	n. 浏览器
surfing the Internet		网上冲浪
E-mail	['i:meil]	n. 电子邮件
BBS（Bulletin Board System）		电子布告栏系统
Blog（Web log）		博客，网络日记
user	['ju:zə]	n. 用户，使用者
anonymous	[ə'nɔniməs]	adj. 无名的，不具名，匿名的
directory	[di'rektəri]	n.（计算机文件或程序的）目录
current	['kʌrənt]	adj. 现在的，现行的，当前的
transfer	[træns'fə:]	vt.& vi. 转移；迁移；传输
complete	[kəm'plit]	adj. 完整的，完全的　vt. 完成，结束
telnet	['telnet]	n. 远程登录
port	[pɔ:t]	n.（计算机与其他设备的）接口；端口，插口
group	[gru:p]	n. 组，群，团体，类，批，簇
logged in		注册登记，登录

Part Ⅰ　Information Query and Publish

WWW is a huge repository of information, anyone can publish Web queries or information. 腾讯官方网站如图 11-1 所示。

CHAPTER 11 About Internet（关于 Internet）

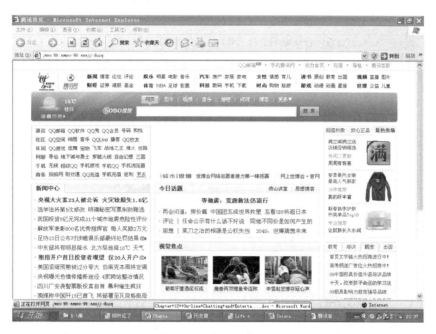

图 11-1 腾讯官方网站

ICQ 官方网站如图 11-2 所示。

图 11-2 ICQ 官方网站

Part Ⅱ Communication for Information

ICQ 网站提供 E-mail 和 Blog 服务，如图 11-3 所示。

图 11-3 ICQ 网站提供 E-mail 和 Blog 服务

1. E-mail

ICQ 邮箱登录界面如图 11-4 所示。

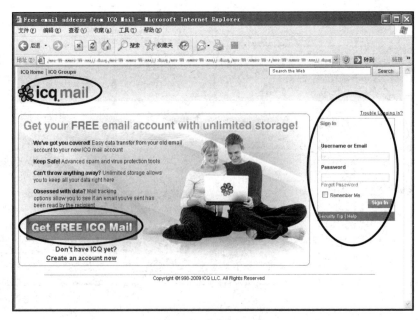

图 11-4 ICQ 邮箱登录界面

CHAPTER 11　About Internet（关于 Internet）　119

> icq.mail ICQ
  邮箱。
> Username or Email
  用户名或邮箱名。
> Password
  密码。
> Forgot Password
  忘记密码。
> Remember Me
  记住密码。
> Sign In
  登录。
> Get FREE ICQ Mail
  获得免费 ICQ 邮箱。

2. Blog

Blog is a weblog.
ICQ Blog 频道如图 11-5 所示。

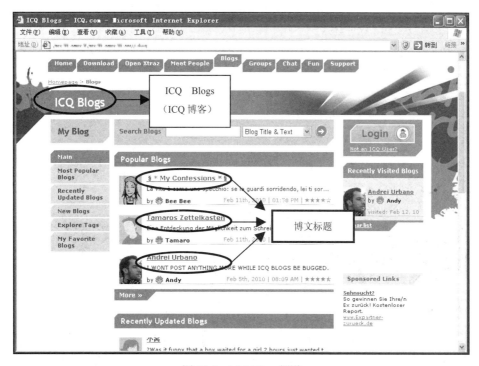

图 11-5　ICQ Blog 频道

3. BBS

BBS is the eletronic bulletin board.
CHINA DAILY BBS 界面如图 11-6 所示。

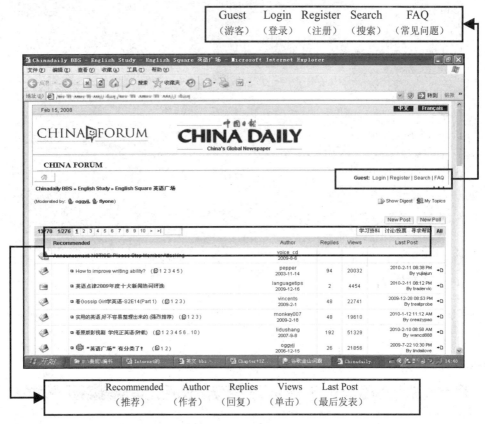

图 11-6 CHINA DAILY BBS 界面

Part Ⅲ Resource Sharing

1. Telnet

远程登录操作界面如图 11-7 所示。

（a）准备远程登录

（b）已经登录到远程计算机

图 11-7 远程登录操作界面

2. FTP

FTP is characterized by fast, informative, transmission of date can be any type of file.
FTP 操作界面如图 11-8 所示。

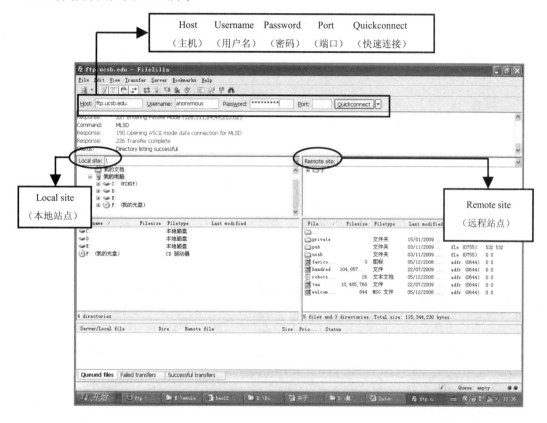

图 11-8　FTP 操作界面

Lesson 2　Search Engine（搜索引擎）

Interrelated Knowledge

　　Internet 是一个巨大的信息资源宝库，所有的 Internet 使用者都希望该宝库中的资源越来越丰富。Internet 中的信息以惊人的速度增长，每天都有新的主机被连接到 Internet 上，每天都有新的信息资源被增加到 Internet 中。由于 Internet 中的信息资源分散在无数台主机之中，故想通过访问每一台主机来获取自己需要的信息显然是不现实的，搜索引擎由此应运而生。
　　如表 11-1 所示是常用的搜索引擎。

表 11-1　常用的搜索引擎

谷歌	百度	雅虎
搜狗	爱问	搜狐

❀ Words and Expressions

search　　　　[sə:tʃ]　　　v. 搜索；搜寻，搜查；调查　　n. 搜寻，探究
engine　　　　['endʒin]　　n. 发动机，引擎
key word　　　　　　　　　关键字

下面举例说明如何搜索国际奥委会的官方网站，如图 11-9～图 11-14 所示。

图 11-9　谷歌搜索主页

CHAPTER 11　About Internet（关于 Internet） 123

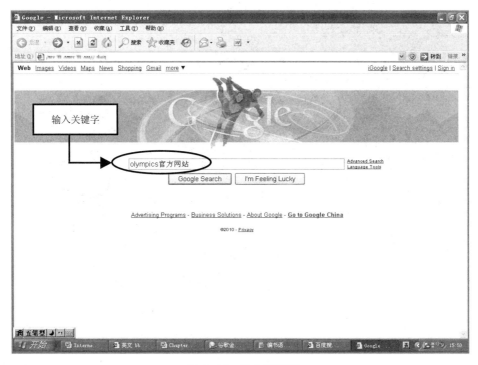

图 11-10　输入关键字

图 11-11　搜索结果显示

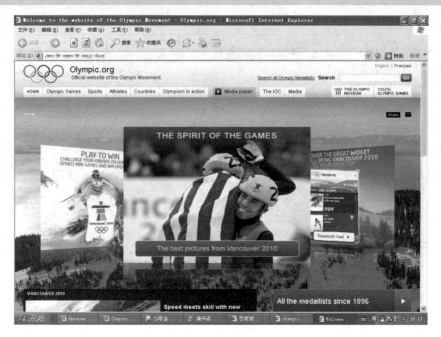

图 11-12　进入国际奥委会网站

图 11-13　国际奥委会网站 Sports 频道

Exercises

1. Translation.

在图 11-13 中，你能识别的奥运会项目有哪些？

项目名称（英文）	项目名称（中文）	项目名称（英文）	项目名称（中文）

2. Thinking.

（1）在图 11-14 中，位于上方的四幅图告诉了我们什么？

图 11-14　国际奥委会网站的 The IOC 频道

（2）搜索引擎的使用有许多小技巧，你知道多少？与同学们一起分享。

参 考 译 文

随着计算机技术的不断发展，"地球村"已不再是一个遥不可及的梦想。我们可以通过 Internet 获取我们想要的各种信息。Internet 是一个全球性的巨大的计算机网络体系，它把全球数万个计算机网络、数千万台主机连接起来，包含了难以计数的信息资源，向全世界提供信息服务，如收发电子邮件，和朋友聊天，进行网上购物，观看影片片断，阅读网上杂志，还可以聆听音乐会，并实现资源共享。Internet 是计算机网络发展的必然产物。

CHAPTER 12

On-line Chatting and Entertainment
（网上聊天与娱乐）

Since the development of the Internet, Online chatting such as Instant Messaging has become more and more popular. People can make friends with others from all over the world online. ICQ is the earliest IM computer program, first developed in 1996. The name ICQ is a homophone for the phrase "I seek you". At present, the most popular IM program in China is QQ. Do you own a QQ number? Add me as a friend with No. 52329396.

Lesson 1　Instant Messaging
（即时通信）

❀ Interrelated Knowledge

Instant Messaging 的中文翻译为"即时通信"，可缩写为 IM，它是一种使人们能在网上识别在线用户并与他们实时交换消息的技术，被很多人称为电子邮件发明以来最酷的在线通信方式。

ICQ 是最早出现的即时通信软件之一，它源自以色列特拉维夫的 Mirabilis 公司（成立于 1996 年 7 月），由几个以色列青年在 1996 年 11 月发明。ICQ 意指"I seek you"，中文意思是"我找你"，俗称网络寻呼机，可以及时传送文字信息、语音信息、聊天和发送文件。当今网络上流行的即时通信软件有腾讯 QQ、MSN Messenger、新浪 UC、Skype、网易泡泡等。

QQ 也就是 OICQ，是腾讯的即时聊天工具。它最初是在模仿 ICQ，即在 ICQ 前加了一个字母 O，意为 opening I seek you，意思是"开放的 ICQ"，但是遭到了侵权控诉，于是就把 OICQ 改为 QQ，即现在我们常用的 QQ。

MSN 全称为 Microsoft Service Network，即微软网络服务。MSN Messenger 是微软开发的即时通信软件，MSN Messenger 网络是一个出自微软的实时通信网络。

❀ Words and Expressions

online	['ɔnlain]	adj. 联机的，在线的
chat	[tʃæt]	v. 聊天，闲谈
development	[di'veləpmənt]	n. 发展，开发区，开发
instant	['instənt]	adj. 立即的，即时的
message	['mesidʒ]	v. 传递信息，通信
popular	['pɔpjulə]	adj. 大众的，流行的，有销路的
develop	[di'veləp]	vt. 发展，开发，冲洗照片
homophone	['hɔməfəun]	n. 同音异形异义字
phrase	[freiz]	n. 短语，习语
seek	[si:k]	v. 寻找
at present		adv. 现在，目前
download	['daunləud]	v. 下载
select	[si'lekt]	vt. 挑选
language	['læŋgwidʒ]	n. 语言
accept	[ək'sept]	vt. 接受，同意，承担（责任等）
agreement	[ə'gri:mənt]	n. 同意，一致，协议
installation	[ˌinstə'leiʃən]	n. 安装，装置

CHAPTER 12 On-line Chatting and Entertainment（网上聊天与娱乐） 129

custom	['kʌstəm]	n. 自定义，自主，习惯，风俗，海关
homepage		主页，通过 Web 进行信息查询的起始信息页
toolbar	['tu:lbɑ:]	n. 工具条（栏）
icon	['aikɔn]	n.【计算机】图标，图符
desktop	['desktɔp]	n. 桌面，台式计算机
account	[ə'kaunt]	n. 账目，报告，估计
password	['pɑ:swə:d]	n. 口令，密码
contact	['kɔntækt]	n. 接触，联系；联系人
nickname	['nikneim]	n. 绰号，昵称
sign in		登录，（使）签到，（使）登记

Part Ⅰ ICQ

1. Download ICQ

下载 ICQ 的操作过程如图 12-1 所示。

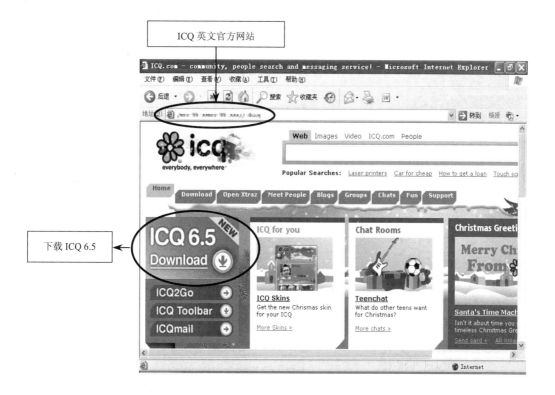

图 12-1 下载 ICQ

2. Install

安装 ICQ 的过程如图 12-2～图 12-5 所示。

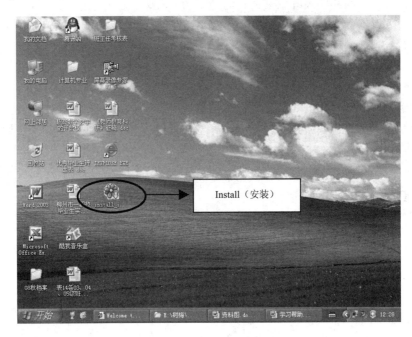

图 12-2　安装 ICQ 的过程 1

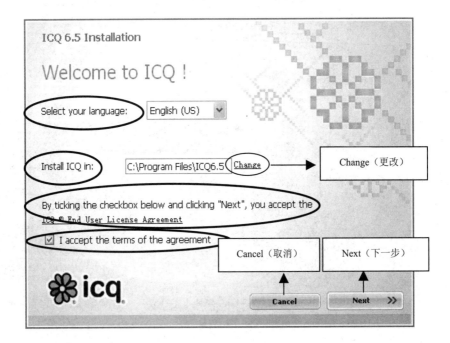

图 12-3　安装 ICQ 的过程 2

CHAPTER 12　On-line Chatting and Entertainment（网上聊天与娱乐）

➢ Select your language
请选择语言。
➢ Install ICQ in
（选择）ICQ 的安装位置。
➢ By ticking the checkbox below and clicking "Next", you accept the <u>ICQ End User License Agreement</u>
勾选以下复选框并单击"下一步"按钮，即表示您接受 ICQ 的最终用户许可协议。
➢ I accept the terms of the agreement
我接受协议条款。

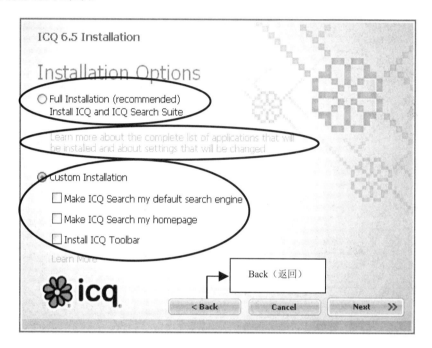

图 12-4　安装 ICQ 的过程 3

➢ Installation options
安装选项。
➢ ○Full Installation (recommended)
完整安装（推荐使用）。
➢ Install ICQ and ICQ Search Suite
安装 ICQ 及 ICQ 的搜索套件。
➢ Learn more about the complete list of applications that will be installed and about settings that will be changed
了解更多关于待安装应用程序完整列表的信息，以及即将更改的设置。
➢ ⊙Custom Installation
　自定义安装。

☐ Make ICQ Search my default search engine
将"ICQ 搜索"设置为我的默认搜索引擎。

☐ Make ICQ Search my homepage
将"ICQ 搜索"设置为我的主页。

☐ Install ICQ Toolbar
安装 ICQ 工具栏。

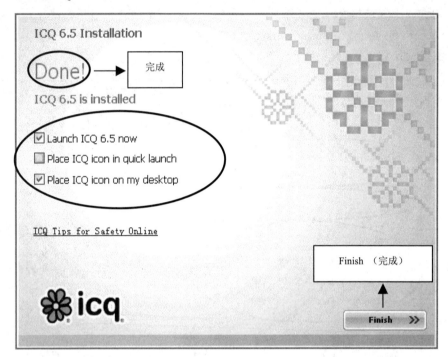

图 12-5　安装 ICQ 的过程 4

➢ ICQ 6.5 is installed
ICQ6.5 已经安装。

☑ Launch ICQ 6.5 now
立即启动 ICQ6.5。

☐ Place ICQ icon in quick launch
将 ICQ 图标放在快速启动栏中。

☑ Place ICQ icon on my desktop
将 ICQ 图标放在桌面上。

➢ ICQ Tips for Safety Online
ICQ 联机安全提示。

3. Login

ICQ 的注册及登录如图 12-6 所示。

CHAPTER 12　On-line Chatting and Entertainment（网上聊天与娱乐）

Account
账户。

- Get a new account
  新用户在此注册。
- Password
  密码。
- Forgot my password
  忘记了密码。
- Save my password
  保存密码。
- Auto sign-In
  自动登录。
- Open on startup
  启动时打开。
- Sign in
  登录。

图 12-6　ICQ 的注册及登录

4. Add contacts

添加联系人的过程如图 12-7 和图 12-8 所示。

- Contact
  联系人。
- Add Contacts
  添加联系人。
- View History
  查看历史记录。
- Show/Hide
  显示/隐藏。
- Sort By
  排序方式。
- Sort Groups
  排序组。
- Search for
  搜索（可按以下不同选项进行搜索）。
- ICQ#
  ICQ 号。
- Email
  电子邮箱。

图 12-7　添加联系人的过程 1

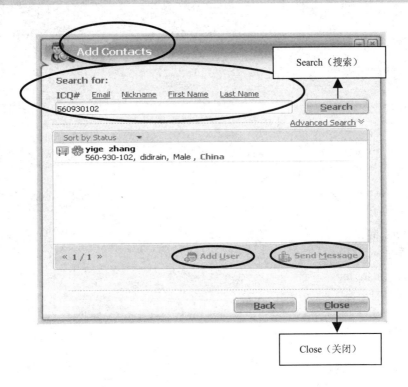

图 12-8　添加联系人的过程 2

> Nickname
> 昵称。
> First Name
> 名字。
> Last Name
> 姓氏。
> Add User
> 添加用户。
> Send Message
> 发送消息。

5. Send message

发送信息的界面如图 12-9 所示。

Part Ⅱ　MSN & QQ

1. QQ

QQ 登录及界面如图 12-10 和图 12-11 所示。

CHAPTER 12 On-line Chatting and Entertainment（网上聊天与娱乐）

图 12-9 发送信息

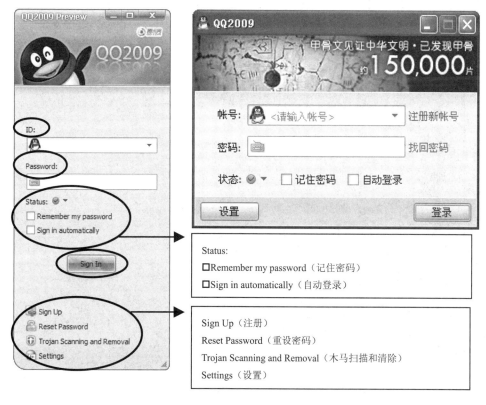

图 12-10 QQ 登录——中英文版界面对照

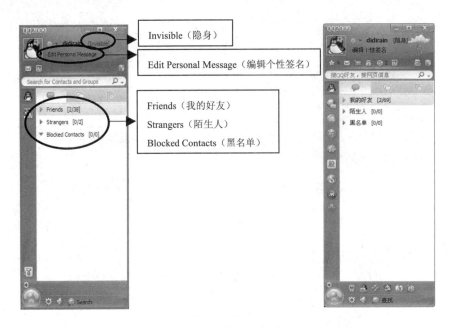

图 12-11　QQ 界面——中英文版对照

2. MSN

MSN Messenger 的界面如图 12-12 所示。

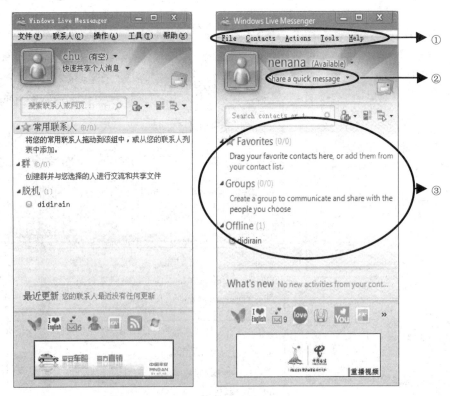

图 12-12　MSN Messenger 的界面——中英文版对照

CHAPTER 12　On-line Chatting and Entertainment（网上聊天与娱乐）

➤ ①　File　Contacts　Actions　Tools　Help

➤ ②　Share a quick message

➤ ③
▲ Favorites
　Drag your favorite contacts here, or add them from your contact list.
▲ Groups
　Create a group to communicate and share with the people you choose
▲ Offline

Lesson 2　Entertainment on the Internet（网上娱乐）

Interrelated Knowledge

　　互联网正飞速发展着，人们也越来越多地感受到互联网对生活的影响及其起到的巨大作用。互联网改变了我们的生活方式，更重要的是互联网正在改变着我们的思维方式，如"百度一下，你就知道"！

　　网上娱乐是网络应用极其重要的一部分，如看小说、电影，听音乐，玩网游，上社交网和玩社交游戏，论坛灌水，使用即时通信软件聊天，看新闻，逛博客等。

Words and Expressions

Warcraft	['wɔːkrɑːft]	n. 魔兽争霸（Blizzard Entertainment 出品的即时战略游戏），军用飞机（战略和战术）
login	[lɔg'in]	登录，注册
file	[fail]	n.【计算机】文件
undo	['ʌn'duː]	v. 取消，解开，松开
redo	[riː'duː]	重做
score	[skɔː]	vi. 得分，记分，得胜
marketplace	['mɑːkitpleis]	n. 集会场所，市场，商场
supplier	[sə'plaiə]	n. 供应者，供应厂商，供应国
trade	[treid]	vi. 做生意，购物，交换
platform	['plætfɔːm]	n. 平台，月台，讲台，坛，计划
update	[ʌp'deit]	n. 更新，补充最新资料

Part Ⅰ　Games on the Internet

"魔兽世界"的登录界面如图 12-13 所示。

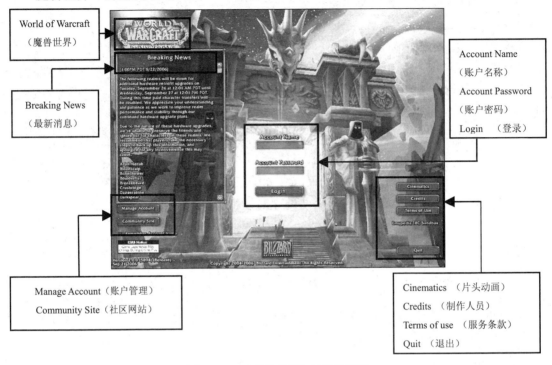

图 12-13　"魔兽世界"的登录界面

"纸牌游戏"的界面如图 12-14 所示。

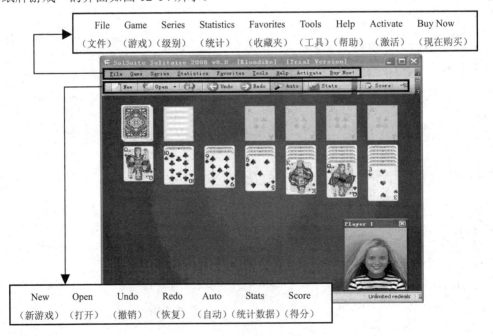

图 12-14　"纸牌游戏"的界面

CHAPTER 12 On-line Chatting and Entertainment（网上聊天与娱乐）

Part Ⅱ Shopping On the Internet

Alibaba.com is one of the world's largest online B2B marketplaces, connecting millions of buyers and suppliers worldwide every day.

阿里巴巴英文官方网站如图 12-15 所示。

图 12-15 阿里巴巴的英文官方网站

Ebay is the other one of a world's largest online trading platform.

易趣的英文官方网站如图 12-16 所示。

图 12-16 易趣的英文官方网站

Part Ⅲ Web TV

网络电视如图 12-17 所示。

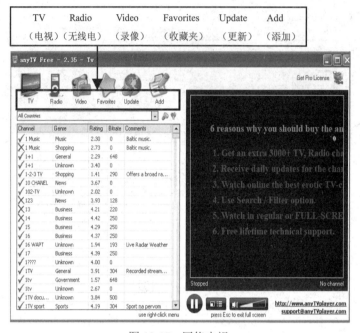

图 12-17 网络电视

Exercises

1. Translation.

在如图 12-18～图 12-23 所示的 QQ2009 的安装步骤中，你能告诉同学们其中关键的提示信息的意思吗？

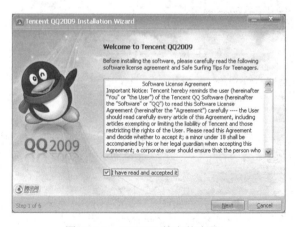

图 12-18 QQ2009 的安装步骤一

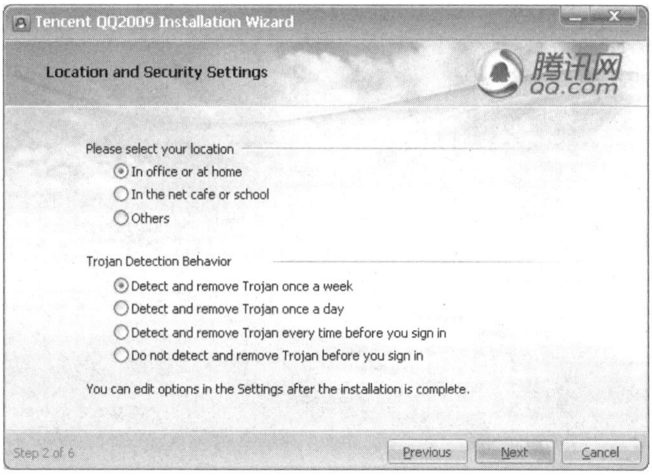

CHAPTER 12　On-line Chatting and Entertainment（网上聊天与娱乐）

图 12-19　QQ2009 的安装步骤二

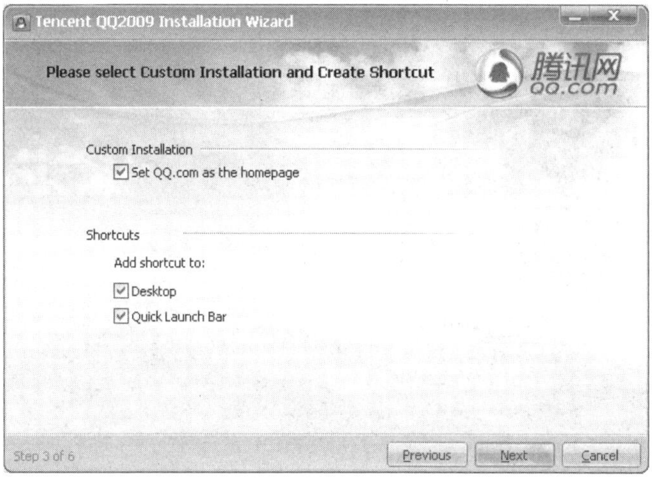

图 12-20　QQ2009 的安装步骤三

图 12-21　QQ2009 的安装步骤四

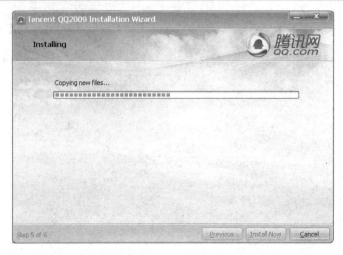

图 12-22　QQ2009 的安装步骤五

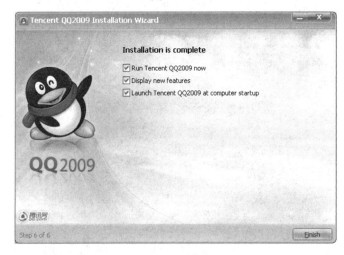

图 12-23　QQ2009 的安装步骤六

2．Thinking．

你知道什么是 Windows Live 吗？与同学们一起分享。

参 考 译 文

　　随着互联网的发展，类似于即时通信的在线聊天已经越来越普及，人们可以在线和世界上任何一个角落的人交朋友。ICQ 就是于 1996 年最早开发出来的即时聊天软件。ICQ 是英文"I Seek You"的谐音，中文意思是我找你。目前在中国最流行的即时聊天软件是 QQ。你有 QQ 号码了吗？加我为好友吧！我的 QQ 号是 52329396。

Appendix A

商 务 篇

基本商务英语

1. 面试英语对话

Interviewer（面试主考官）	Interviewee（应试者）
Can you sell yourself in two minutes? （你能在两分钟内自我推荐一下吗？）	I feel I am a hardworking, responsible and diligent person in any project. （我觉得我是一个对所从事的每一个工作项目都很努力、负责、用功的人。）
Give me a summary of your current job description. （给我一个你对目前工作的概括性的说明。）	I have been working as a secretary in a company for five years. I try my best to do everything. （我在一家公司做秘书已经五年了。我尽力做好每一件事。）
What contribution did you make to your previous organization? （你对之前的工作单位有何贡献？）	I have finished many projects well, and I am sure I can apply my experience to this position. （我很好地完成了多项工作任务，我相信我能将我的经验用在这份工作上。）
What do you think you are worth to us? （你认为你对我们有何价值呢？）	I feel I can make some positive contributions to your company in the future. （我觉得我对贵公司能作出积极的贡献。）
What is most important in your life right now? （眼下你生活中最重要的是什么？）	To secure employment hopefully with your company. （能在贵公司任职对我来说最重要。）
What is your strongest traits? （你个性上最大的特点是什么？）	Helpfulness, caring and adaptability. （乐于助人、关心他人和较强的适应能力。）
What is important to you in a job? （在你的工作中最重要的是什么？）	Challenge, the feeling of accomplishment, and knowing that you have made a contribution. （挑战和成就感。）
Why do you want to leave your current job? （你为什么要离开目前这份工作？）	There is no room for the career growth and advancement I would like. （那里缺乏一个让我在事业上成长和晋升的空间。）
We have very long working hours. Would you be able to work in the evening? （我们工作时间相当长，你愿意上晚班吗？）	Of course, I will try my best to finish my work. （当然，我会尽我所能做好所有工作。）
Good luck to you! （祝你好运！）	Thanks a lot. （非常感谢。）

2. 工作中的常用句子

In the middle of something? （正在忙吗？）
Please make two copies of this document. （请把这份文件复印两份。）
Would you type this letter? （能否把这封信打出来？）
Would you please call Mr. Black? （你可不可以打电话给布莱克先生？）
I'll put her on the phone. Just a second. （我会请她听电话，请稍等。）
Would you care to leave a message? （你愿意留言吗？）
The meeting begins at nine o'clock. （会议从九点钟开始。）
I hope this meeting is productive. （我希望这是一次富有成效的会谈。）
He's really good at his job. （他工作得心应手。）
It's your job. You should do it by yourself. （这是你的工作，你应该自己做。）

续表

I refuse to work overtime during the weekend. （我拒绝在周末时加班。）	
I want to find a well-paying job. （我想找一份待遇好的工作。）	
I'm thinking of changing jobs. （我想换一份工作。）	
I was referred to you by Mr. Green. （是格林先生介绍我来的。）	
Business is booming. （生意日趋繁荣。）	
There's been extensive marketing research done on this project. （这个计划已做了广泛的市场调查。）	
I'm sure you'll be pleased with this product. （我敢保证你会喜欢这种产品的。）	
Thank you for your coming. （谢谢您的光临。）	

3. 工作、生活中的常用词汇

英语词汇	中文解释	英语词汇	中文解释
typist	打字员	Monday	星期一
receptionist	接待员	Tuesday	星期二
programmer	计算机程序员	Wednesday	星期三
secretary	秘书	Thursday	星期四
policeman	警察	Friday	星期五
director	导演	Saturday	星期六
photographer	摄影师	Sunday	星期日
artists	艺术家	January (Jan.)	一月
painter	画家	February (Feb.)	二月
musician	音乐家	March (Mar.)	三月
singer	歌唱家	April (Apr.)	四月
designer	服装设计师	May	五月
beautician	美容师	June (Jun.)	六月
model	模特	July (Jul.)	七月
stewardess	空中小姐	August (Aug.)	八月
tour guide	导游	September (Sept.)	九月
doctor	医生	October (Oct.)	十月
nurse	护士	November (Nov.)	十一月
public servants	公务员	December (Dec.)	十二月

4. 常见电子商务专业术语

英语	中文
Advertisement on Internet	网上广告
Agent	代理人
Agent's commission	代理费
After-sale service	售后服务
Airport construction fee	机场建设费
Anti-fake label	防伪标志

续表

英　语	中　文
Ask Job on Internet	网上求职
ATM (automated Teller Machine)	自动取款机
Authentication	身份认证
Authentication and Accredit	认证和授权
Bar coding	条形码
BBS(Bulletin Board System)	电子公告栏系统
Bookstore Online	网上书店（如当当网上书店）
B2B(Business to Business)	企业对企业的电子商务（如阿里巴巴）
B2C(Business to Consumer)	企业对消费者的电子商务（如卓越）
B2E(Business to Employee)	企业内部的电子商务（如学校知识管理系统）
B2G(Business to Government)	企业对政府机构的电子商务（如政府机构的采购信息）
Business hours	营业时间
Business lobby	营业厅
Business registration	工商登记
Business registration certificate	工商登记证
CA (Certification Authority)	认证机构
Cash register	收款机
C2C(Consumer to Consumer)	消费者对消费者的电子商务（如淘宝）
CEO(Chief Executive Official)	首席执行官
Certified goods	正品
Choice goods	精品
Clearance sale	甩卖
Commercial speculation	商业炒作
Credit cards	信用卡
Digital Signatures	数字签名
E-bank	电子银行
E-book	电子图书
E-cash	电子现金
E-check	电子支票
E-catalog	电子商品目录
E-Commerce	电子商务
E-commerce platform	电子商务平台
E- Mall	电子购物中心
E- money	电子货币
E-wallet	电子钱包
E-zine	电子杂志
Enterprise Network	企业网
Fortune Global 500	财富全球500强（美国《财富》杂志评选）
Freeware	免费软件
G2C (Government to Consumer)	政府机构对消费者的电子商务（如电子医疗服务）
Investigate on Internet	网上调查

续表

英　语	中　文
IP（Internet Phone）	网络电话
IPRs （Intellectual Property Rights）	知识产权
Logistics	现代物流
Management Group(Team)	管理团队
Netbug	网虫
Netiquette	网络礼仪
Netizen	网民
Netnews	网络新闻
Network Language	网络语言
Newbie	网络新手
Online Marketing	线上营销
Price tag	价格标签
Price war	价格战
Promotion On Internet	网上促销
Report of Business Plan	商业计划书
Sales agent	销售代理商
Service after Sell Online	在线售后服务
Signature	签名
SOHO（Small Office Home Office）	在家办公（代表一种自由、弹性而新型的工作方式）
Talk	对话
Time limit	时间限制
Videoconferencing	视频会议
Workgroups	工作组
Yellow page	黄页（企业名录）

APPENDIX B

常用的翻译工具

无论是你平时浏览网页还是阅读文献都会或多或少遇到几个难懂的外语词汇,这时我们就不免要翻翻词典了。在计算机上使用的词典工具可以分为两种:在线词典,通过访问网站进行查询翻译;离线词典,就是可以不用联网,只要下载安装并运行就可以方便取词翻译了。

1. 离线词典

常用的离线词典如表 B-1 所示。

表 B-1　常用的离线词典

谷歌金山词霸	词霸豆豆	有道桌面词典

谷歌金山词霸合作版是金山与谷歌面向互联网翻译市场联合开发,适用于个人用户的免费翻译软件。它支持中日英三语查询,有取词、查词、查句、全文翻译、网页翻译等功能。	词霸豆豆是通过互联网向广大英语学习用户提供即时、小巧的在线英语查词工具,可通过互联网给用户提供更好的服务。	这是网易最新出品的一款免费软件。其功能全面,不但具有常规的英汉、汉英、英英翻译功能,还能够提供普通字典里所无法收录的各类词汇的网络释义。
灵格斯词霸	雅虎乐译	金山词霸试用版

灵格斯支持全球 60 多个国家语言的互查互译、支持多语种屏幕取词、索引提示和语音朗读,是新一代的词典翻译专家。	雅虎乐译的最大特点是轻巧快捷,不管您到哪个英文网站,都会为您指点迷津,从而带来了最人性化、最智能的翻译体验。	金山词霸是目前使用最为广泛的汉英双语工具。它不仅词汇量大,而且能够即时取词,使用也相当方便。

每种离线词典各有自己的特色，不过，大多数的词典的核心翻译功能大同小异。下面以谷歌金山词霸为例讲述离线词典的用法。

1）词典功能

谷歌金山词霸收录了《现代英汉综合大词典》、《汉英词典》（新）等经典词典，涵盖金山词霸百万余词条。其词典功能非常强大，如图 B-1 所示。

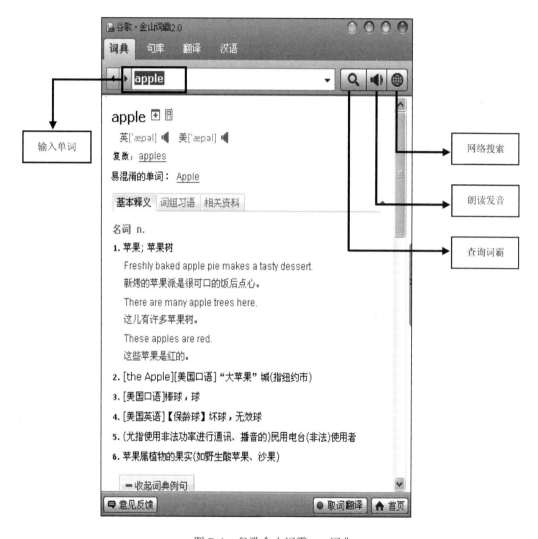

图 B-1　谷歌金山词霸——词典

2）句库功能

谷歌金山词霸收录了 80 万情景例句，覆盖生活情景对话的方方面面！直接输入句子或关键字就可以找到所有相关联的句型和用法，如图 B-2 所示。

3）翻译功能

谷歌金山词霸的翻译功能分为文本翻译和网页翻译。

（1）文本翻译：文本翻译功能即指能直接对文本进行中、英文互译，如图 B-3 所示。

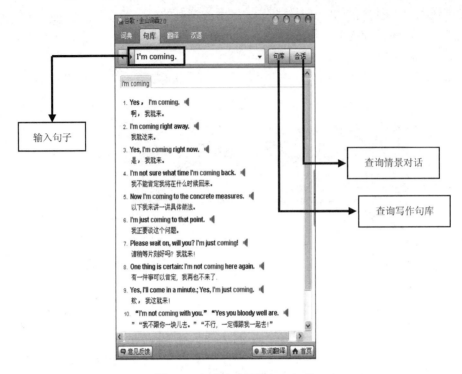

图 B-2　谷歌金山词霸——句库

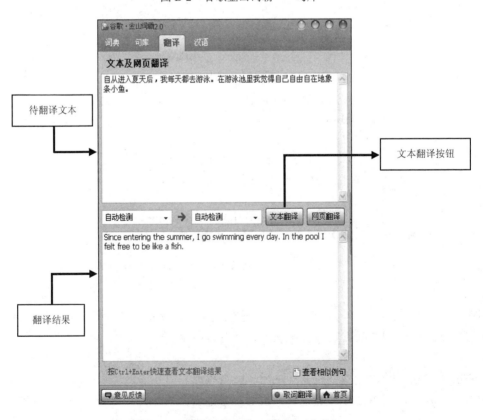

图 B-3　谷歌金山词霸——翻译（文本）

（2）网页翻译：谷歌金山词霸支持网页翻译。直接输入网址并选择语种，即可快速实时地显示翻译结果，如图 B-4 所示。

图 B-4　谷歌金山词霸——翻译（网页）

翻译网页后的 ICQ 官方网站主页如图 B-5 所示。

图 B-5　翻译网页后的 ICQ 官方网站主页

4）汉语功能

谷歌金山词霸的最新特点之一就是加入了全新独立的汉语模块,可以针对输入的汉语词汇查询相关的知识,如图 B-6 所示。

图 B-6　谷歌金山词霸——汉语

5）取词翻译

屏幕取词,即可选任意的单词或词组进行即时翻译,如图 B-7 所示。

屏幕取词翻译示例如图 B-8 所示。

2．在线词典

常用的在线词典如表 B-2 所示。

APPENDIX B 常用的翻译工具 155

图 B-7 谷歌金山词霸——取词翻译

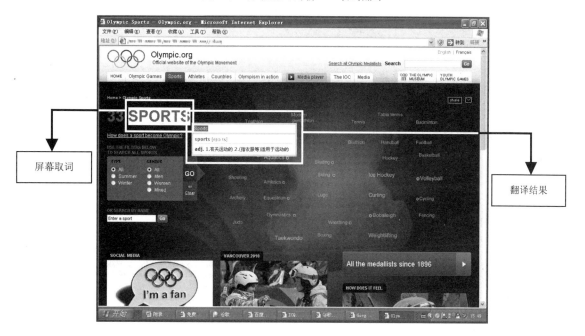

图 B-8 屏幕取词翻译示例

表 B-2 常用的在线词典

金山爱词霸 http://www.iciba.com/	海词在线词霸 http://dict.cn/	译典通 http://www.dreye.com.cn/
它是目前最好的线上词典工具之一。其词汇量涵盖了 150 余本词典辞书，70 余个专业领域，28 种常备资料，中、日、英网际大辞海，可提供在线及时更新，能第一时间掌握流行词汇表。	海词在线词典由中国留学生范剑淼创建。它正式使用于 2003 年 11 月 27 日。虽然它的词汇量没有爱词霸庞大，但是它提供了大量例句并配有真人发音，可以帮助矫正发音问题。	网站的所有者为英业达(上海)有限公司。译典通提供了较大量的词汇并配有真人发音，同时可以查询同义词/反义词，词形变化等。另外它还可以用来查询日语词汇，并配有日语、英语学堂。
星际译王 http://www.stardict.cn/	有道词典 http://dict.youdao.com/	洪恩在线 http://study.hongen.com/dict/
它提供了大量的词汇翻译，支持多语言翻译。不过要使用它的全部功能必须注册成为会员，注册会员是免费的。星际译王还提供了 Firefox 插件，您可以试试看。	有道是网易自主研发的搜索引擎，并提供词典功能。有道词典的释义也是来自译典通，但又有很多创新，如英中翻译、网络解释、例句查询等，还可以创建自己的单词本。	输入英文，可以查询英文常用词义、词根、词缀、词性、特殊形式、详细解释与例句、同义词、反义词、相关短语等；输入中文，可以查询对应英文单词。

离线词典和在线词典的用法大致相同，下面是海词在线的使用说明。

1）海词在线的主界面

海词在线的主界面如图 B-9 所示。

图 B-9 海词在线主界面

2）输入待译单词

海词在线的应用示例 1 如图 B-10 所示。

APPENDIX B 常用的翻译工具　157

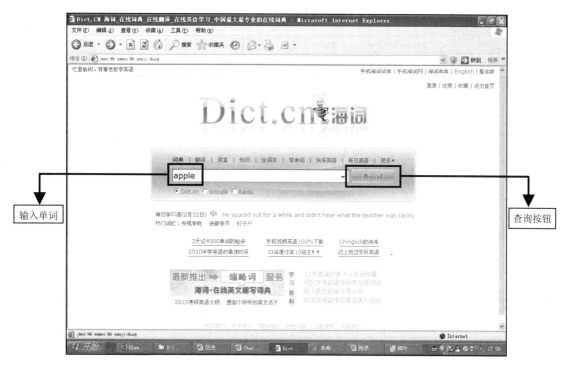

图 B-10　海词在线的应用示例 1

3）显示翻译结果

海词在线的应用示例 2 如图 B-11 所示。

图 B-11　海词在线的应用示例 2

APPENDIX C

计算机词汇的特征

词汇是语言系统中结构最松散的部分，因此也最容易起变化。词汇系统是个发展的开放性系统，它不仅广纳对其有用的词汇，还在自身各类词汇间的相互渗透、派生和转变中不断丰富和扩大自己。计算机英语词汇通过派生法构成的新词语的频繁出现，是计算机英语词汇最显著的特征之一。美国《新闻周刊》就有一个栏目叫"cyberspace"，这是美国人用以代替electronic-world 的词，它因小说家威廉•吉布森 1984 年的小说 Neuromancer 而家喻户晓。它起源于数学家诺伯特•韦特在 1948 年创造的新词 cybernetics（控制论）。将两词相比较，就会发现 cyberspace 一词是借用了 cybernetics 中的 cyber（电子的、电脑的）这一词根（root）加上 space 而构成的。与此词根有关的用在计算机世界里的词随处可见，如 cyber-announce，cyber-age，cyber-revolution，cyberland，cyber-culture，cyberphilia，cyberphobia，cyberslut 等。从词汇学的角度分析，语言中最小的"语音语义的结合体"称为"词素"，在词中有明确的语义，表达主要的意义。计算机英语词汇的词素一般都有一个含义比较明确的语义，如上所述的 cyber 表示"电子的"或"电脑的"。

1. 派生词（Derivation）

派生形式具有构词力强、词义宽广和结合灵活等特点，为计算机英语词汇提供了取之不尽、用之不竭的源泉。这类词汇非常多，它是根据已有的词加上某种前后缀，或以词根生成，或以构词成分形成新的词。科技英语词汇中很大一部分来源于拉丁语、希腊语等外来词，有的是直接借用，有的是在它们之上不断创造出新的词汇。这些词汇的构词成分（前、后缀，词根等）较固定，构成新词以后便于读者揣度词义，易于记忆。

1）前缀

采纳前缀构成的单词在计算机专业英语中占了很大比例，通过下面的实例可以了解这些常用的前缀构成词。

①multi-	多功能	②hyper-	超级	③super-	超级
multiprogram	多道程序	hypercube	超立方	superhighway	超级公路
multimedia	多媒体	hypercard	超级卡片	superpipeline	超流水线
multiprocessor	多处理器	hypermedia	超媒体	superscalar	超标量
multiplex	多路复用	hypertext	超文本	superset	超集
multiprotocol	多协议	hyperswitch	超级交换机	superclass	超类
④inter-	相互、在……间	⑤micro-	微型	⑥tele-	远程
interface	接口、界面	microprocessor	微处理器	telephone	电话
interface	隔行扫描	microkernel	微内核	teletext	图文电视
interlock	联锁	microcode	微代码	telemarketing	电话购物
internet	互联网（因特网）	microkids	微机迷	telecommuting	家庭办公
interconnection	互联	microchannel	微通道	teleconference	远程会议

单词前缀还有很多，其构成可以同义而不同源（如拉西、希腊），可以互换，如：

multi, poly	相当于 many	如 multimedia, polytechnique
uni, mono	相当于 single	如 unicode, monochrome
bi, di	相当于 twice	如 bichloride, dichloride
equi, iso	相当于 equal	如 equality, isograph
simili, homo	相当于 same	如 similarity, homogeneous

| semi,hemi | 相当于 half | 如 semiconductor,hemicycle |
| hyper,super | 相当于 over | 如 hypertext,superscalar |

2）后缀

后缀是在单词后部跟上构词结构，形成新的单词，如：

① -scope 探测仪器　　　　② -meter 计量仪器　　　　③ -graph 记录仪器

baroscope	验压器	barometer	气压表	barograph	气压记录仪
telescope	望远镜	telemeter	测距仪	telegraph	电报
spectroscope	分光镜	spectrometer	分光仪	spectrograph	分光摄像仪

④ -able 可能的　　　　　　⑤ -ware 件（部件）　　　　⑥ -ity 性质

enable	允许、使能	hardware	硬件	reliability	可靠性
disable	禁止、不能	software	软件	availability	可用性
programmable	可编程	firmware	固件	accountability	可核查性
portable	便携的	groupware	组件	integrity	完整性
scalable	可缩放的	freeware	赠件	confidentiality	保密性

2. 复合词

复合词（Compound）是科技英语中的另一类词汇，通常分为复合名词、复合形容词和复合动词等。复合词通常以小横杠"-"连接单词构成，或者采用短语构成。随着复合词的进一步发展，有的去掉了小横杠，并经过缩略成为另一个词类，即混成词。

复合词的实例如下所示。

① -based 基于，以……为基础　　　　② -centric 以……为中心的

rate-based	基于速率的	client-centric	以客户为中心的
credit-based	基于信誉的	user-centric	以用户为中心的
file-based	基于文件的	host-centered	以主机为中心的
Windows-based	以 Windows 为基础的		

③ -oriented 面向……的　　　　　　④ -free 自由的、无关的

object-oriented	面向对象的	lead-free	无线的
market-oriented	市场导向	jumper-free	无跳线的
process-oriented	面向进程的	paper-free	无纸的
thread-oriented	面向线程的	charge-free	免费的

⑤ info- 信息、与信息有关的

info-channel	信息通道
info-tree	信息树
info-world	信息世界
info-sec	信息安全

3. 赋新

词的意义是人们的知识对客观事物、观点和关系的反映，是在词中固定下来的某种联想。通过这种联想，词可以成为语言外部的客观事物，表达指物性意义。计算机词汇中吸收了许多学科领域中的尖端科技成果，创造了大量的新词，经历了语义变化及隐喻修饰之后，原词的意

义已荡然无存，以全新的语义出现，如下所示。

Firewall	防火墙	pop-up	弹出
Gopher	地鼠	fanin	扇入
Laptop	手提电脑（膝上型微机）	onboard	在板
Desktop	桌面，单机，小型机	pushup	拉高
Videotape	录像带	download	下载
Online	在线	login	登录
Logout	注销	placeholder	占位
Database	数据库	point-and-click	点击
point-to-point	点到点	drag-and-drop	拖放
plug-and-play	即插即用	line-by-line	逐行
easy-to-use	易用的	store-and-forward	存储转发
off-the-shelf	现成的	operator-controllable	操作员控制的
peer-to-peer	对等的	over-hyped	过度宣扬的
leading-edge	领先的	front-user	前端用户
end-user	最终用户	sign-off	取消
sign-on	登录	sound-blaster	声霸
pull-down	下拉	diskette-only	磁盘专有的

4．混成词（Blending）

混成词不论在公共英语还是科技英语中都大量出现，也有人将它们称为缩合词（与缩略词区别）或融会词，它们多是名词，有的地方也将其作为动词用，对这类词汇可以通过其构词规律和词素进行理解。这类词汇将两个单词的前部拼接、前后拼接或者将一个单词前部与另一词拼接构成新的词汇，如"modem"翻译成"调制解调器"，该词借用 modulate（改变波的幅度、频率或相位）的前三个字母 mod 与 demodulate（把已改变的波的幅度、频率或相位恢复为原来的状态）的前三个字母 dem，两个字母 d 合为一个，成为新词 modem，准确地反映了 modulate 和 demodulate 两者的功能特征。其他的例子有如下所示。

deltree（delete+tree）	删除子目录
smartdrv（smart+drive）	智能驱动器
motel（motor+hotel）	汽车旅馆
brunch（breakfast+lunch）	早中饭
smog（smoke+fog）	烟雾
codec（coder+decoder）	编码译码器
compuser（computer+user）	计算机用户
transeiver（transmitter+receiver）	收发机
mechatronic（meachanical+electronic）	机械电子学
calputer（calculator+computer）	计算式电脑
syscall（system+call）	系统调用
signoded（sign+code）	厂家名称和代号

5. 缩略词（shortening）

在高速发展的信息时代，人们不堪忍受那些特别冗长繁杂的词语，于是使用简明的新词语取而代之。缩略词在文章索引、前序、摘要、文摘、电报和说明书等科技文章中频繁采用。对计算机专业来说，在程序语句、程序注释、软件文档和文件描述中也采用了大量的缩略词作为标识符、名称等。缩略词的出现方便了印刷、书写、速记及口语交流等，但也增加了阅读和理解的难度。人们常常把两个或两个以上的词缩略，只用其首字母组成缩略词，以期用最少的词汇来清楚地表达更多的信息量，如 CPU 是 Central Processing Unit（中央处理单元，中央处理器）的首字母缩写词。另一种缩略词是将较长的英语单词取其首部或者主干构成的与原词同义的短单词。

缩略词也可能有形同而义异的情况，如果遇到这种情况，翻译时应当根据上下文确定词义，并在括号内给出其原型组合词汇。缩略词可以分为如下几种。

1）压缩和省略

将某些太长、难拼难记、使用频繁的单词压缩成一个短小的单词，或取其头部、或取其关键音节，这类截短词（clipped word）使用起来方便快捷、简洁高效，特别是在菜单、提示和系统命令中，更是能起到以简代繁的效果，例如：

  dir（directory）   列目录
  del（delete）   删除
  flu（influenza）   流感
  lab（laboratory）   实验室
  auto（automobile）   汽车
  math（mathematics）   数学
  iff（ if only if）   当且仅当

截短词中还有一种有趣的现象，即把单词中的元音字母剔除，只保留辅音字母，在书写或印刷时一般用黑体或斜体，以便于区分，例如：

  drv（drive）   驱动器
  hlp（help）   帮助
  ctrl（control）   控制键

2）缩写（Acronym）

将某些词组和单词集合中每个实意单词的第一或者首部几个字母重新组合，组成为一个新的词汇，作为专用词汇使用。在应用中它形成了两种类型，如下所示。

（1）通常以小写字母出现，并作为常规单词。

  radar（radio detecting and ranging，无线电探测与定位）雷达
  laser（light amplification by stimulated emission of radiation，受激辐射式光频放大器）激光器
  sonar（sound navigation and ranging，声波搜索与定位）声呐
  spool（simultaneous peripheral operation on line，同时外围设备在线操作）假脱机

（2）以大写字母出现，具有主体发音音节。采用首字母缩略词可以使计算机英语词汇使用起来方便、简捷。它们易于上口，节省书写和打印时间及文章篇幅等，因此这类词汇增长神速，应用十分广泛，在计算机世界中占有相当大的比例，且十分活跃。

  PC（Personal Computer）个人计算机，微机
  UPS（Uninterrupted Power Supply）不间断电源

DOS（Disk Operating System）磁盘操作系统
CAI（Computer Aided Instructions）计算机辅助教学
BASIC（Beginner's All-purpose Symbolic Instruction Code）初学者全能符号指令代码，一种高级程序设计语言
FORTRAN（FORmula TRANsaltion）公式翻译，一种高级程序设计语言
COBOL（Common Business Oriented Language）面向商务的通用语言
RISC（Reduced Instruction Set Computer）精简指令集计算机
CISC（Complex Instruction Set Computer）复杂指令集计算机

（注：本节选自 张政. 新编计算机英语教程. 北京：电子工业出版社，2004）

APPENDIX D

单词表

A

abacus	['æbəkəs]	n. 算盘	(1)
absolute	['æbsəlu:t]	adj. 绝对的，完全的	(1)
accept	[ək'sept]	vt. 接受，同意，承担（责任等）	(12)
accessory	[æk'sesəri]	adj. 附属的（副的，辅助的） n. 附件	(9)
account	[ə'kaunt]	n. 账目，报告，估计	(12)
accuracy	['ækjurəsi]	n. 精确（性），精确（程度），准确（性）	(1)
action	['ækʃən]	n. 动作，行动	(8)
advanced	[əd'vɑ:nst]	adj. 高级的，先进的	(3)
agreement	[ə'gri:mənt]	n. 同意，一致，协议	(12)
album	['ælbəm]	n. 相册，集邮册	(9)
anonymous	[ə'nɑnəməs]	adj. 无名的，不具名，匿名的	(11)
anti	['ænti]	adj. 反对的	(9)
application	[,æpli'keiʃən]	n. 应用；应用软件程序	(6)
authenticity	[ɔ:θen'tisiti]	n. 确实性，真实性	(9)
auto	['ɔ:təu]	pref. 自动的，自己的	(5)
average	['ævəridʒ]	n. 平均数，平均水平	(8)

B

backdoor	['bækdɔ:]	a. 秘密的	(9)
background	['bækgraund]	n. 背景，幕后	(8)
beep	[bi:p]	n. 哔哔声	(4)
binder	['baində]	n. 捆绑机	(9)
block	[blɔk]	n. 阻塞（物），障碍（物） v. 阻塞	(9)
Browser	['brauzə]	n. 浏览器	(11)
burn	[bə:n]	v. 燃烧，烧着 n. 燃烧	(9)
bus	[bʌs]	n.【计】总线	(10)
business	['biznis]	n. 商业，生意，事务	(3)

C

cable	['keibl]	n. 电缆	(10)
calculate	['kælkjuleit]	v. 计算，估计，核算，计划，认为	(1)
capture	['kæptʃə]	vt. 捕获，（用照片等）留存	(9)
chairman	['tʃɛəmən]	n. 主席，会长 vt. 担任……的主席（议长）	(2)
champion	['tʃæmpjən]	n. 冠军，优胜者 vt. 保卫，拥护	(1)
chat	[tʃæt]	v. 聊天，闲谈	(12)
check	[tʃek]	v. 检查，阻止，核对 n. 检查，支票，账单	(4)
christen	['krisn]	vt. 为……命名	(7)
commercialize	[kə'mə:ʃəlaiz]	vt. 使商业化	(7)
common	['kɔmən]	adj. 常见的，普遍的，共同的	(4)
communicate	[kə'mju:nikeit]	v. 交流，传达，沟通	(6)

单词	音标	释义	章节
company	['kʌmpəni]	n. 同伴，客人，一群，连队，公司	(3)
compare	[kəm'pɛə]	v. 比较，比喻，对照	(7)
compatible	[kəm'pætəbl]	adj. 能共处的，可并立的，适合的，兼容的	(3)
compilation	[kɔmpi'leiʃən]	n. 编译，编辑	(9)
compile	[kəm'pail]	vt. 编译，编纂	(9)
complete	[kəm'plit]	adj. 完整的，完全的　vt. 完成，结束	(11)
component	[kəm'pəunənt]	n. 元件，组成部分，成分　adj. 组成的，构成的	(4)
compress	[kəm'pres]	vt. 压缩，压榨	(6)
configuration	[kənˌfigju'reiʃne]	n. 结构，形状，【计算机】配置	(5)
configure	[kən'figə]	v. 配置	(5)
connect	[kə'nekt]	v. 连接，接通	(10)
consist	[kən'sist]	vi. 由……组成，构成，在于，符合	(4)
consume	[kən'sju:m]	v. 消耗，花费，挥霍	(1)
contact	['kɔntækt]	n. 接触，联系；联系人	(12)
contain	[kən'tein]	vt. 包含，容纳	(6)
convert	['kɔnvə:t]	v. 变换，（使）转变	(9)
copyright	['kɔpirait]	n. 版权，著作权　adj. 版权的	(4)
core	[kɔ:]	n. 核心，要点	(4)
corporation	[ˌkɔ:pə'reiʃne]	n. 公司，法人，集团	(2)
corrupt	[kə'rʌpt]	adj. 有缺陷的；有错误的 v. 引起……错误；破坏	(6)
count	[kaunt]	v. 计算，视为，依赖　n. 计数，总数	(8)
current	['kʌrənt]	adj. 现在的，现行的，当前的	(11)
custom	['kʌstəm]	n. 自定义，自主，习惯，风俗，海关	(12)

D

单词	音标	释义	章节
decompress	[di:kəm'pres]	vt. 缓缓排除压力，减压	(9)
default	[di'fɔ:lt]	n. 假设值，默认（值） v. 默认，【计算机】缺省	(5)
delay	[di'lei]	v.推迟，延误　n.推迟，延期	(9)
design	[di'zain]	n. 设计，图样；vt. 想象，设计，计划	(3)
desktop	['desktɔp]	n. 桌面，台式电脑	(12)
destination	[desti'neiʃən]	n. 目的地，终点	(9)
develop	[di'veləp]	vt. 发展，开发，冲洗照片	(12)
development	[di'veləpmənt]	n. 发展，开发区，开发	(12)
device	[di'vais]	n. 装置，设计，策略，设备	(3)
diagnosis	[daiəg'nəusis]	n. 诊断	(7)
digital	['didʒitl]	adj. 数字的，数码的，手指的，电子的	(1)
directory	[di'rektəri]	n.（计算机文件或程序的）目录	(11)
disable	[dis'eibl]	v. 使……失去能力，【计算机】禁用	(5)
distance	['distəns]	n. 距离	(10)

download	['daunləud]	v. 下载	(12)

E

efficiently	[i'fiʃəntli]	adv. 有效地	(6)
elapse	[i'læps]	v. 逝去，过去	(6)
electric	[i'lektrik]	adj. 电的，令人激动的，鲜亮的	(3)
E-mail	['i:meil]	n. 电子邮件	(11)
embed	[im'bed]	vt. 使插入，使嵌入　vi. 嵌入	(8)
enable	[i'neibl]	vt. 使……能够，【计算机】启用，激活	(5)
engine	['endʒin]	n. 发动机，引擎	(11)
entertainment	[ˌentə'teinmənt]	n. 娱乐	(9)
entrepreneur	[ˌɔntrəprə'nə:]	n. 企业家，主办者，承包商	(2)
entry	['entri]	n. 进入，入口，登记，条目	(6)
error	['erə]	n. 错误，过失，谬误，误差	(4)
Ethernet	['i:θənet]	n. 以太网	(6)
executive	[ig'zekjutiv]	n. 执行者，主管，行政部门	(2)
exit	['ekzit, sit]	v. 退出，离去　n. 出口，退场	(5)
extraction	[iks'trækʃən]	n. 抽出，取出，抽出物	(9)

F

familiar	[fə'miljə]	adj. 熟悉的，熟知的	(8)
famous	['feiməs]	adj. 著名的，一流的	(3)
favorite	['feivərit]	adj. 喜爱的　n. 最喜爱的人或物，收藏夹	(9)
fiber	['faibə]	n. 纤维（物质），力量	(10)
figure	['figə]	n. 图形，数字，形状；人物，外形，体型	(2)
file	[fail]	n. 【计算机】文件	(12)
filter	['filtə]	n. 筛选　v. 过滤，渗透	(9)
firewall	['faiəwɔ:l]	n. 防火墙	(9)
flash	[flæʃ]	n. 闪光，闪现，计算机动画技术	(3)
folder	['fəuldə]	n. 文件夹，折叠者	(9)
formally	['fɔ:məli]	adv. 正式地，形式上地	(1)
format	['fɔ:mæt]	n. 格式	(9)
found	[faund]	v. 建立，创立，创办	(2)
function	['fʌŋkʃən]	n. 功能，函数	(5)
function	['fʌŋkʃən]	n. 功能，函数　vi. 运行，起作用	(8)

G

group	[gru:p]	n. 组，群，团体，类，批，簇	(11)

H

hacker	['hækə]	n. 计算机黑客	(9)
hardware	['hɑ:dwɛə]	n.（计算机）硬件	(10)
hint	[hint]	n. 暗示	(7)
homophone	['hɔməfəun]	n. 同音异形异义字	(12)

单词	音标	释义	章节
hyperlink	['haipəlink]	n. 超链接	(8)
I			
icon	['aikɔn]	n.【计算机】图标，图符	(12)
illustrator	['iləstreitə]	n. 插图画家	(3)
image	['imidʒ]	n. 图像，影像 vt. 想象，描绘，反映	(6)
include	[in'klu:d]	vt. 包括，包含	(4)
industry	['indəstri]	n. 工业，产业，勤勉	(2)
information	[ˌinfə'meiʃən]	n. 信息，情报，新闻，资料，询问	(2)
insert	[in'sə:t]	v. 插入，嵌入 n. 插入物	(8)
inspect	[in'spekt]	vt. 调查，检阅	(6)
inspector	[in'spektə]	n. 检查员，巡视员	(8)
install	[in'stɔ:l]	vt. 安装，安置	(6)
installation	[ˌinstə'leiʃən]	n. 安装，装置	(12)
instant	['instənt]	adj. 立即的，即时的	(12)
integrated	['intigreitid]	adj. 整合的，综合的，集成的	(3)
interface	['intəfeis]	n. 界面，接口 v. 连接，作接口	(7)
international	[ˌintə'næʃənəl]	adj. 国际的，世界性的 n. 国际比赛	(3)
Internet	['intə:net]	n. 因特网，国际互联网	(3)
J			
joke	[dʒəuk]	n. 笑话，玩笑	(9)
K			
keyboard	['ki:bɔ:d]	n. 键盘；vt. 用键盘输入	(3)
L			
language	['læŋgwidʒ]	n. 语言	(12)
launch	[lɔ:ntʃ]	v. 发射，发动，发起	(9)
leading	['li:diŋ]	adj. 领导的，主要的，在前的	(2)
login	[lɔg'in]	登录，注册	(12)
logo	['ləugəu]	n. 图形，商标，标识语	(3)
M			
macro	['mækrəu]	n. 宏	(9)
mainboard	['meinbɔ:d]	n. 主机板，主板	(3)
maintenance	['meintinəns]	n. 维护，保持，维修	(6)
marketplace	['mɑ:kitpleis]	n. 集会场所，市场，商场	(12)
marquee	[mɑ:'ki:]	n. 大天幕，华盖	(8)
memory	['meməri]	n. 记忆，内存，回忆，【计算机】存储	(3)
message	['mesidʒ]	v. 传递信息，通信	(12)
miscellaneous	[misi'leinjəs]	n. 杂货，杂项	(9)
modem	['məudem]	n. 调制解调器	(10)
monitor	['mɔnitə]	n. 班长，【计算机】显示器，监视器 v. 监视，监听，监督	(3)

单词	音标	释义	页码
mouse	[maus]	n. 老鼠，鼠标，胆小如鼠的人	(3)
multimedia	[ˌmʌltiˈmiːdiə]	adj. 多媒体的 n. 多媒体	(6)

N

network	[ˈnetwəːk]	n. 网络；vt. 联络，交流	(10)
nickname	[ˈnikneim]	n. 绰号，昵称	(12)
normal	[ˈnɔːməl]	n. 常态，标准 adj. 正常的，正规的	(5)
notify	[ˈnəutifai]	v. 通知，通告，报告	(4)
number	[ˈnʌmbə]	n. 号码，数字 vi. 总计，编号 vt. 编号	(8)

O

online	[ˈɔnlain]	adj. 联机的，在线的	(12)
operate	[ˈɔpəreit]	vi. 操作，运转	(7)
optical	[ˈɔptikəl]	adj. 光学的，视觉的	(10)
Oracle	[ˈɔːrəkl]	Oracle 是世界领先的信息管理软件开发商；殷墟出土的甲骨文（oracle bone inscriptions）的英文翻译的第一个单词	(2)
outpost	[ˈautpəust]	n. 前哨，前哨基地，警戒部队	(9)

P

panel	[ˈpænl]	n. 面板，嵌板，仪表盘 v. 嵌镶	(6)
paragraph	[ˈpærəgrɑːf]	n. 段落，节 vt. 将……分段	(8)
parameter	[pəˈræmitə]	n. 参数，参量，决定因素	(7)
partition	[pɑːˈtiʃən]	vt. 区分，隔开，分割 n. 分割，隔离物	(6)
password	[ˈpɑːswəːd]	n. 口令，密码	(12)
pending	[ˈpendiŋ]	adj. 待定的，即将发生或来临的	(9)
percent	[pəˈsent]	n. 百分比 adv. 百分之…… adj. 百分之……的	(6)
personage	[ˈpəːsənidʒ]	n. （历史、小说、戏剧中的）人物、角色；名流	(2)
personal	[ˈpəːsənl]	adj. 私人的，个人的	(3)
phrase	[freiz]	n. 短语，习语	(12)
platform	[ˈplætfɔːm]	n. 平台，月台，讲台，坛，计划	(12)
popular	[ˈpɔpjulə]	adj. 大众的，流行的，有销路的	(12)
port	[pɔːt]	n. （计算机与其他设备的）接口；端口，插口	(11)
precision	[priˈsiʒən]	n. 精确度，准确（性） adj. 精确的；准确的	(1)
premiere	[ˈpremiɛə]	v. 初次公演，初演主角	(3)
press	[pres]	v. 按，压	(5)
primarily	[praiˈmərili]	adv. 首先，主要地	(1)
printer	[ˈprintə]	n. 打印机，印刷工	(3)
priority	[praiˈɔriti]	n. 优先权，优先顺序，优先	(5)
profile	[ˈprəufail]	n. 概要	(9)
program	[ˈprəugræm]	n. 程序，计划 vt. 编制程序，拟……计划	(6)
prompt	[prɔmpt]	vi. 提示 n. 提示，提示的内容	(9)
property	[ˈprɔpəti]	n. 性质；财产	(8)

单词	音标	释义	章节
protection	[prə'tekʃən]	n. 保护，防护	(9)
provider	[prə'vaidə]	n. 供应者，赡养者	(2)
R			
reader	['ri:də]	n. 读者，读物，读本	(3)
redo	[ri:'du:]	重做	(12)
registry	['redʒistri]	n. 注册，登记，登记簿	(6)
remove	[ri'mu:v]	v. 消除，除去，脱掉，搬迁	(6)
repeater	[ri'pi:tə]	n. 【计】中继器，转发器	(10)
replacement	[ri'pleismənt]	n. 更换，接替者	(1)
reset	['ri:'set]	v. 重新设定，重新放置，将……恢复原位	(4)
revolution	[ˌrevə'lu:ʃən]	n. 革命，旋转，转数	(2)
ring	[riŋ]	n. 环，戒指	(10)
root	[ru:t]	n. 根，根源，根本	(8)
S			
score	[skɔ:]	vi. 得分，记分，得胜	(12)
screen	[skri:n]	n. 屏，幕，银幕	(5)
script	[skript]	n. 脚本，原稿，手稿，手迹	(8)
scroll	[skrəul]	n. 卷轴，目录	(9)
search	[sə:tʃ]	v. 搜索；搜寻，搜查；调查 n. 搜寻，探究	(11)
security	[si'kju:riti]	n. 安全；防护措施	(9)
seek	[si:k]	v. 寻找	(12)
select	[si'lekt]	vt. 挑选	(12)
sequence	['si:kwəns]	n. 顺序，次序，序列 vt. 按顺序排好	(5)
series	['siəri:z]	n. 系列，连续，丛书	(4)
server	['sə:və]	n. 服务器	(10)
setup	['setʌp]	n. 设置，安装，计划	(5)
shortcut	['ʃɔ:tkʌt]	n. 捷径；快捷方式，快捷键	(6)
similar	['similə]	adj. 类似的，同样的	(5)
sort	[sɔ:t]	n. 种类 v. 分类，整理，使明确	(9)
split	[split]	v. 分离，分开	(9)
statistics	[stə'tistiks]	n. 统计，统计数字，统计学	(6)
status	['steitəs]	n. 地位，身份，情形，状况	(6)
subfolder	['sʌbfəuldə]	n. 子文件夹	(9)
sum	[sʌm]	n. 总数，金额 v. 总计，概括	(8)
super	['sju:pə]	adj. 超级的，极好的 n. 主管人，负责人	(1)
supervisor	['sju:pəvaizə]	n. 监督人，管理人 【计算机】（网络）超级用户	(5)
supplier	[sə'plaiə]	n. 供应者，供应厂商，供应国	(12)
support	[sə'pɔ:t]	n. 支持，援助，供养 vt. 支援，帮助，支持	(5)
switch	[switʃ]	v. 转换，改变，交换 n. 【计】转换器	(7)

单词	音标	释义	章节
system	['sistəm]	n. 系统，体系，制度	(5)

T

单词	音标	释义	章节
tag	[tæg]	n. 标签，附属物	(8)
technologically	[teknə'lɔdʒikli]	adv. 技术上地	(7)
technology	[tek'nɔlədʒi]	n. 技术，工艺（学）	(2)
telnet	['telnet]	n. 远程登录	(11)
terminology	[ˌtə:mi'nɔlədʒi]	n. 术语，术语学	(10)
tip	[tip]	n. 提示	(7)
title	['taitl]	n. 头衔，名称，标题，所有权，资格，冠军	(2)
toolbar	['tu:lbɑ:]	n. 工具条（栏）	(12)
track	[træk]	n. 跑道，轨道，踪迹； v. 跟踪，追	(9)
trade	[treid]	vi. 做生意，购物，交换	(12)
transfer	[træns'fə:]	vt.& vi. 转移；迁移；传输	(11)
transistor	[træn'zistə]	n. 晶体管，三极管	(1)
treaty	['tri:ti]	n. 条约，协定	(6)
Trojan	['trəudʒən]	n. 特洛伊	(9)

U

单词	音标	释义	章节
undo	[ʌn'du:]	v. 取消，解开，松开	(12)
unprecedented	[ʌn'presidəntid]	adj. 空前的，前所未有的	(1)
update	[ʌp'deit]	n. 更新，补充最新资料	(12)
User	['ju:zə]	n. 用户，使用者	(11)

V

单词	音标	释义	章节
various	['vɛəriəs]	adj. 各种各样的	(9)
version	['və:ʃən]	n. 版本，说法	(7)
virus	['vaiərəs]	n. 病毒	(9)
volume	['vɔljum]	n. 卷，册，容量，音量	(8)

W

单词	音标	释义	章节
Warcraft	['wɔ:krɑ:ft]	n. 魔兽争霸（Blizzard Entertainment 出品的即时战略游戏），军用飞机（战略和战术）	(12)
warning	['wɔ:niŋ]	n. 警告，警报	(5)
Web	[web]	n. 网页	(11)
wizard	['wizəd]	n. 向导；有特殊才干的人，奇才；术士	(6)
workstation	['wə:k.steiʃən]	n. 工作站	(7)
worm	[wə:m]	n. 虫，蠕虫	(9)

注：后面括号的数字代表该单词所在的章节。

参 考 文 献

[1] 张政. 新编计算机英语教程. 北京：电子工业出版社，2004.
[2] 陈俊宇，辛燕清. 计算机专业英语. 北京：冶金工业出版社，2006.
[3] 比尔·金兰顿. 软件英语基础教程. 北京：电子工业出版社，2007.
[4] 王泰. 计算机专业英语. 北京：高等教育出版社，2003.